# About Island Press

Since 1984, the nonprofit Island Press has been stimulating, shaping, and communicating the ideas that are essential for solving environmental problems worldwide. With more than 800 titles in print and some 40 new releases each year, we are the nation's leading publisher on environmental issues. We identify innovative thinkers and emerging trends in the environmental field. We work with world-renowned experts and authors to develop cross-disciplinary solutions to environmental challenges.

Island Press designs and implements coordinated book publication campaigns in order to communicate our critical messages in print, in person, and online using the latest technologies, programs, and the media. Our goal: to reach targeted audiences—scientists, policymakers, environmental advocates, the media, and concerned citizens—who can and will take action to protect the plants and animals that enrich our world, the ecosystems we need to survive, the water we drink, and the air we breathe.

Island Press gratefully acknowledges the support of its work by the Agua Fund, Inc., The Margaret A. Cargill Foundation, Betsy and Jesse Fink Foundation, The William and Flora Hewlett Foundation, The Kresge Foundation, The Forrest and Frances Lattner Foundation, The Andrew W. Mellon Foundation, The Curtis and Edith Munson Foundation, The Overbrook Foundation, The David and Lucile Packard Foundation, The Summit Foundation, Trust for Architectural Easements, The Winslow Foundation, and other generous donors.

The opinions expressed in this book are those of the author(s) and do not necessarily reflect the views of our donors.

# ECOLOGICAL RESTORATION

# Ecological Restoration

*Principles, Values, and Structure of an Emerging Profession*

**SECOND EDITION**

Andre F. Clewell and James Aronson

Washington | Covelo | London

Library of Congress Cataloging-in-Publication Data

Clewell, Andre F.
  Ecological restoration : principles, values, and structure of an emerging profession / Andre F. Clewell and James Aronson. – 2nd ed.
      p. cm.
  Includes bibliographical references and index.
  ISBN 978-1-61091-167-2 (cloth : alk. paper) – ISBN 1-61091-167-9 (cloth : alk. paper) – ISBN 978-1-61091-168-9 (pbk. : alk. paper) – ISBN 1-61091-168-7 (pbk. : alk. paper) 1. Restoration ecology. I. Aronson, James, 1953- II. Society for Ecological Restoration International. III. Title.
  QH541.15.R45C54 2012
  577–dc23

                                                                                    2012026960

 Printed on recycled, acid-free paper

Manufactured in the United States of America
10 9 8 7 6 5 4 3 2 1

*Keywords:* Island Press, coadapted species, ecological restoration, restoration ecology, restoration practitioner, Society of Ecological Restoration, ecological impairment, ecological recovery, disturbance regime, reference model, professional certification, virtual field trip, holistic restoration, cultural ecosystem, ecological attributes, ecological impairment, ecological process, ecological processes, ecological recovery, ecological restoration, ecological trajectory, ecological wholeness, ecosystem services, historic continuity, reference model, reference site, restoration brazil, Restoration Chile, Restoration Florida, Restoration France, Restoration India, Restoration Oregon, restoration profession, Restoration South Africa, restoration values, restoring natural capital, self-sustainability

# CONTENTS

*To*

John Cairns Jr.

In recognition of his outstanding career as a dedicated and persistent advocate of ecological restoration and other kinds of environmental improvement, and for his penchant to grasp quickly the detrimental consequences of ecosystem impairment to human well-being and to explain them compellingly to his contemporaries. His early recognition of the importance of restoring natural capital was a substantial advancement for our discipline. Much of what we say in this book was said succinctly by John twenty-five years ago or more. However, the world has yet to take his wisdom to heart, and we feel compelled to repeat it.

*Between my finger and my thumb*
*The squat pen rests.*
*I'll dig with it.*
–Seamus Heaney*

The first edition of this book was a landmark advance in the struggle to apply clear ecological principles to the practice of restoration, for the benefit of the practitioner in the field. Conversely, and equally necessary, the book also included a series of lively bulletins from the field to the desk and the laboratory; they reminded scientists that principles have to be constantly adjusted to reflect new and often case-specific encounters with reality.

This second edition fine tunes that process significantly, but it remains—as all good books in the dynamic territory of restoration must—a work in progress.

That is what lends it a very particular excitement. You can hear the sound of a new, globally significant science and profession being freshly excavated in these pages. Andy Clewell and James Aronson are digging for the rich layers of meaning, the multiple implications for practice, that underlie surface perceptions of that seductively hopeful phrase, *ecological restoration*.

In his early poem "Digging," Seamus Heaney envied his forebears for their practical, useful skill with a sharp spade on a turf bog, and yet he finally asserts that the writer also makes a real impact and can reveal the roots beneath the topsoil of our world.

A similar tension between the claims of practice and the claims of theory gives energy to this book. Clewell and Aronson are exceptionally well equipped to express it, because each embodies it in their daily life. They are scientists with wet boots, with the authentic texture of the Earth on their pages.

Longer ago now than he probably cares to remember, Clewell left a successful university career in botany and ecology. He morphed into an even more success-

---

* **"Digging," from** *Death of a Naturalist*

ful entrepreneur—and pioneer—in ecological restoration, while continuing to publish, rooting his restless theorizing in projects he has worked on personally.

Aronson remains wedded to the academy but, as a passionate advocate of the restoration of natural capital, he spends as much time debating ecological (and social) issues with practitioners on the ground as he does teasing out the more abstract frontiers of restoration science.

But why do I say that this extensively revised second edition of their already authoritative "yellow book" must remain a work in progress? Surely it must be something close to definitive at this stage?

Well, No, and No, for two reasons: first, restoration remains a teenager among the grizzled veterans of environmental sciences and strategies. True, its stature and potential are increasingly widely recognized. A special section on the field in *Science*, published in 2009, midway between the edition of this book that you are holding now and the first one, pronounced this remarkable accolade at the outset: "Our planet's future may depend on the maturation of the young discipline of ecological restoration."

But its relatively recent emergence means that its core principles are still in process of formation. And even when they appear to find a definitive expression— the key example being *The SER Primer on Ecological Restoration* of 2004, of which Clewell and Aronson were, with the late Keith Winterhalder, the principal authors—the ink is barely dry before revisions are called for. And indeed, Aronson and others were working on a new version of the primer even as he was finishing this book.

Second, some of this constant ferment is due to ongoing deep philosophical differences, though they are not always recognized as such, between different schools and individuals laying claim to the restoration label.

The authors identify three such schools: a "legacy" model that stresses restoration to a favored past state, often conceived of as both pristine and static; a "utility" model that stresses the restoration of ecosystem services of immediate and obvious benefit to humans, like potable water and fertile soil; and a "recovery" model, which they themselves favor.

As they define it, this recovery model has much to recommend it. An ecosystem restored according to this model "will develop complexity, self-organization, and resilience, but its future expressions of biodiversity will not necessarily emulate prior states."

The recovery model shares the legacy model's insight that prior states of a degraded ecosystem offer invaluable clues to its potential trajectory into the future. But it eschews any romantic temptation to attempt a return to an idealized past.

The recovery model also shares the utility model's insight that human societies will only pay for the restoration of ecosystems that provide them with services; but it extends the concept of services beyond the immediate and obvious, to include biodiversity—both as a value in itself, and as the provider of many services of which we are currently unaware—and the aesthetic and spiritual values we derive from restored ecosystems.

The authors do well to clearly identify these models, and to attempt to dispel the confusion that arises from a well-meaning tendency to paper over divisions within the restoration movement: "There is a tendency by different parties and people with interests in ecological restoration to advocate one model and ignore the existence of the other two, as if the field of ecological restoration were unified. No unity exists. In reality, ecological restoration as a maturing discipline is rife with growing pains. We all need to acknowledge that other models of ecological restoration exist besides the one to which we adhere. We need to search for common understanding and shared values."

It is important to stress, however, that the field of restoration is in constant flux, not just because it is young and more than a little fissiparous philosophically, but also because of both the scale, and the particularity, of the challenge that every restoration project involves.

Each of the three restoration models demands our engagement with an equation where many, and usually most, of the variables are unknown to us. We are not working in a closed laboratory when we set out on a restoration project. To restore even the simplest ecosystem would—ideally—require experimentally tested knowledge of a web of relationships in a completely open system and, for all practical purposes, an infinite one.

And what do we really know about these systems? Well, any soil ecologist will tell you that we have not even identified many of the species of microflora and microfauna that form, literally and metaphorically, the base of any ecosystem; much less have we grasped their interactions and functions.

You could be forgiven for thinking that the case is very different for plants and animals. And yet, as the authors bluntly tell us, the assumption that ecologists know the roles that a given species of plant or animal plays in an ecosystem "cannot be readily confirmed with confidence in the field."

Small wonder, then, for all these reasons, that it is proving extraordinarily difficult to establish general principles that can guide all restoration everywhere. Even at the most local scale, every single project will have unique features, and will have to be approached as case specific to a considerable degree, if there is to be much chance of success.

And at the global scale, restoration in different biomes demands radically different strategies. As the authors point out, the origins of the restoration movement in the American Midwest, and more generally in temperate boreal regions, inevitably conditioned the principles developed by its foundational thinkers.

Even on their home ground, these principles, based on overly simplistic concepts derived from Clementsian natural succession, have been under challenge for decades. As Steve Hopper and others have demonstrated, their application elsewhere on the globe, where climate and soil conditions are often utterly different, and sometimes unique, is likely to be more misleading still.

Does all this add up to a characterization of ecological restoration as a hopelessly quixotic enterprise, a scandalous waste of scarce conservation resources in pursuit of a delusion?

On the contrary, this book demonstrates, cogently and accessibly, a great deal of evidence to the contrary. While very few restoration projects have met all their targets, many have met enough of them to demonstrate decisively that, as a strategy to combat environmental degradation, restoration is indeed becoming the valued weapon of choice for many communities and organizations.

Those of us who attempt to grasp the vital opportunity that restoration offers owe much to pioneers like Clewell and Aronson. Their digging with both spade and pen has opened up clear and stimulating pathways to this renewal of our engagement with the natural world.

*Paddy Woodworth*

In this book we offer a comprehensive account of the practice of ecological restoration, the discipline that recovers impaired ecosystems and returns them to wholeness. We wrote it for all those who initiate, finance, administer, manage, plan, and, most particularly, implement ecological restoration projects. We also had in mind all of those who serve in supportive roles, such as growers who provide nursery stock, ecologists who monitor and evaluate projects, social scientists who provide liaison with the public sector, and agency personnel who perform regulatory functions. Above all, we wrote this book for students and young professionals from around the world who are considering careers in ecological restoration, so that they may understand the challenges and appreciate the profound satisfaction that this discipline will bring them.

This is not a book on ecology per se, although we describe at length the scientific underpinnings of restoration practice in what we hope is both a substantive and accessible manner. In addition, we examine in depth the many values that ecological restoration fulfills. We discuss how ecological restoration projects are structured and administered. We also explore allied fields of natural resource management and conservation with which ecological restoration is sometimes conflated.

We did not write the book to attract a broad public readership, but those members of the general public with an interest in our field will hopefully find much stimulating and provocative material. We hope to attract readers who are public policy makers and business leaders whose professional activities and decisions regarding environmental issues will benefit from a thorough understanding of our field. We also hope to attract environmental philosophers and those writers who are attempting to satisfy a rapidly growing public interest in ecological restoration. We want to reach this diverse audience because its understanding of restoration is crucially important in order for our field to realize its full potential. Indeed, ecological restoration is emerging as a meeting ground for many different fields and interest groups in all strata of society and in all cultures around the world. A common understanding of its precepts and its practice is essential for effective dialogue and collaboration.

Those who are familiar with the first edition of this book (Clewell and Aronson 2007) will find that this, the second edition, has been revised and reorganized from top to bottom. The impetus for revision was to improve what reviewers had already proclaimed to be a successful book, update it regarding new developments in this rapidly advancing discipline, and add new topics for discussion. We solicited review comments concerning the first edition and incorporated suggestions as best we could. Our insertion of case histories of restoration projects by those practitioners who conducted them, which we called Virtual Field Trips (VFTs), was highly popular in the first edition. We replace them with entirely new Virtual Field Trips here for added interest. This edition lacks the appendix that appeared in the first edition, which was a verbatim copy of a foundation document of the Society for Ecological Restoration (SER) entitled *Guidelines for Developing and Managing Ecological Restoration Projects*. That document is available on SER's web page (www.ser.org), and it was important to include it in the first edition, because it had not been published previously in hard copy.

The first edition was written in large part as an elaboration on another foundation document, *The SER Primer on Ecological Restoration* (SER 2004). We continue to refer to the *SER Primer*, but this edition is more of a stand-alone volume than was the first edition. Versions of the *SER Primer* were issued in 2002 and 2004. The text is identical in both versions, and the only difference lies in their formatting. In our first edition (Clewell and Aronson 2007), we cited SER (2002). In this edition, we cite SER (2004), the version that is currently posted on the SER website. Readers should be aware that both the *SER Guidelines* and the *SER Primer* will probably undergo review soon for possible updating and revision.

We have prepared a somewhat revised glossary for this edition, covering the perplexing array of terms that relate to ecological restoration. Definitions are tailored to the way we use those terms in this book. For the most part, we retained the same definitions as in the first edition; however, we added more terms and updated others, relying primarily on definitions accepted by **The Economics of Ecosystems and Biodiversity** (TEEB 2010) and van Andel and Aronson (2012).

## Layout of Sections and Chapters

The book is not a "how-to" manual that instructs the reader in particular tactics and methods for performing ecological restoration at a given project site, or in a given ecoregion. Instead, it attempts to cover all of the *other* topics relevant to restoration practice. The book consists of twelve chapters, arranged in four parts.

Part I, entitled *Why We Restore*, begins with an overview of the entire field of ecological restoration, what it means to civilization, and introductions of important principles that recur throughout the book. Next, we consider the personal, cultural, socioeconomic, and ecological values that are fulfilled by ecosystems after they have largely recovered from impairment. We offer a conceptual model that organizes these values coherently. The section ends by addressing the globally relevant issue of ecosystem impairment.

Part II, entitled *What We Restore*, examines the complex topic of ecological recovery, the meaning of which is central to the definition of ecological restoration, and how recovery is accomplished. We describe ecological attributes that give restored ecosystems wholeness when they are attained. The controversial topic of the restoration of semicultural ecosystems is thoroughly treated.

Part III, entitled *How We Restore*, addresses the essential topic of reference systems and reference models, which inform almost every aspect of restoration planning. Strategic approaches to ecological restoration are examined, including the factors that determine the intensity of effort needed to complete a restoration project. The necessary steps that are common to every ecological restoration project are enumerated, starting with the initial conceptualization of a project and ending with the publication of its case history. Stakeholder involvement is a recurring theme throughout.

Part IV, entitled *Ecological Restoration as a Profession*, examines the relationship of the field to other disciplines, including ecological engineering, landscape design, and the restoration of natural capital (RNC). The section also deals with professional training and certification; the ways that restorationists and other professionals perceive the discipline of ecological restoration; and topically urgent issues for which ecological restoration has relevance, such as climate change. We conclude the book with recommendations that we feel will advance our emerging profession.

## How to Use This Book

Different readers will want different kinds of information from this book. Many will want to read it in sequence. Others may want to read the Virtual Field Trips first as general background before tackling the text. Later chapters are more specialized and may mean more to those who are already experienced restorationists. Most chapters are more or less independent essays and can be read in any order, particularly by those with experience in the field.

We wrote this book to refine definitions, clarify concepts, illuminate current trends, encourage interdisciplinary alliances, and stimulate our readership to develop new visions. We recognize that this book is only a temporal contribution to a rapidly developing discipline. We hope that it will contribute to a global dialogue that pushes ecological restoration forward, in synergy with the related activities of ecological engineering, ecological economics, and sustainability science. We are ready to participate in this dialogue, and to that end we invite your response by email, addressed to us at clewell@verizon.net and james.aronson@cefe.cnrs.fr.

**Acknowledgments:** Tein McDonald and Karen Holl read parts of advanced versions of this edition and made a number of incisive suggestions to clarify and expand our messages. Paddy Woodworth—journalist extraordinaire—examined the book and challenged us to rethink some of its passages. He asked that we acknowledge a conflict of interest that bothered him as he wrote the foreword for this book, since he assisted in its editing. We would share his concern if he had not provided us with constructive criticism that we could only call "tough love." We thank Bérengère Merlot and Christelle Fontaine for tremendous support as research and editorial assistants, and all our numerous colleagues around the world for ongoing debate and dialogue. It is customary for authors to tip their hats to editors of publishing houses who were assigned the task of smoothing rough edges before their books went into production. Instead, we express our gratitude to Barbara Dean of Island Press and her associate Erin Johnson, who assumed much larger roles by insisting by means of their reviews and criticisms that we perform to our highest potentials, both intellectually and as writers.

# PART I

# Why We Restore

Part 1 addresses the imperatives for ecological restoration. Chapter 1 presents an overview of the scope of this book. There we introduce the principles on which ecological restoration is based and the terms and concepts essential to our discourse. In chapter 2, we examine the personal, cultural, socioeconomic, and ecological values that ecological restoration addresses. Without an examination of our motives for performing ecological restoration, we may lack a clear appreciation of what we are actually doing and why it's important, in which case ecological restoration becomes just another way to make a living, or a weekend pastime. In chapter 3, we turn our attention to ecosystems—the object of ecological restoration, and start to distinguish between the degrees of stress and disturbance to which ecosystems can be exposed. We describe the ecological consequences that occur when an ecosystem has been disturbed to the point of *impairment*—one of our key terms—and then requires ecological restoration by restoration practitioners to ensure recovery.

# Chapter 1

# Overview

Ecological restoration is the process of assisting the recovery of an ecosystem that has been degraded, damaged, or destroyed (SER 2004). From an ecological perspective, it is an intentional activity that reinitiates ecological *processes* that were interrupted when an *ecosystem* was impaired. From a conservation perspective, it recovers *biodiversity* in the face of an unprecedented, human-mediated extinction crisis. From a socioeconomic perspective, *ecological restoration* recovers *ecosystem services* from which people benefit. From a cultural perspective, ecological restoration is a way that we strengthen our communities, institutions, and interpersonal relationships by participation in a common pursuit. From a personal perspective, ecological restoration allows us to reconnect with the rest of Nature and restore ourselves as we restore impaired ecosystems. All of these perspectives on ecological restoration distill down to a simple truth: Nature sustains us; therefore, we serve our own interests when we reciprocate and sustain Nature.

While globally cumulative, ecological restoration is necessarily a local endeavor. The decision to restore represents a long-term commitment of land and resources. Ideally, that decision is reached in consensus by all who are affected. A restored ecosystem contributes to peoples' ecological and socioeconomic security and their well-being into the indefinite future. The benefits of ecological restoration are intergenerational. People develop appreciation for local ecosystems when they participate in decisions regarding restoration, and their respect for ecosystems increases if they become actively engaged in restoration activities.

Ecological restoration reinitiates ecological processes, but we cannot intervene and create desired outcomes directly. Instead, we manipulate *biophysical* properties of an impaired ecosystem to facilitate resumption of processes that can only be performed by living organisms. The restoration *practitioner* assists ecosystem recovery much as a physician assists the recovery of a patient. Patients heal themselves under the physician's supervision, care, and encouragement. Similarly, ecosystems respond to assistance provided by restoration practitioners.

Once ecological restoration project activities are completed, a successfully restored ecosystem self-organizes and becomes increasingly self-sustaining in a dynamic sense. It again becomes resilient to *disturbance* and can maintain itself to the same degree as would be expected of an undisturbed ecosystem of the same kind in a similar position in the local *landscape*. In other words, the intent is to recover an impaired ecosystem to a condition of wholeness or intactness. A "whole" ecosystem is characterized by possession of a suite of ecological attributes that are discussed in chapter 5. We use the term *holistic ecological restoration* to distinguish such comprehensive efforts from partial restorative actions that are limited to incremental ecosystem recovery or ecological improvement.

In spite of our ideal to recover an impaired ecosystem to a condition of total *self-sustainability*, the era of Earth's history when intact ecosystems were entirely self-sustainable has come to a close, for two reasons. First, human-mediated environmental *impacts* have become so pervasive globally, and often so severe locally, that many restored ecosystems require ongoing *ecosystem management* to prevent them from slipping into an impaired *state* once again. Second, many seemingly natural ecosystems coevolved with human inhabitants, whose *traditional cultural practices* have transformed them into *semicultural ecosystems*. Such systems degrade from disuse following abandonment and become candidates for ecological restoration. If they are restored to their semicultural state, then cultural practices that previously maintained them should be resumed to ensure their sustainability.

Ecosystems are not static. They evolve in response to natural and *anthropogenic* modifications in the external environment and to internal processes that govern *species composition* and abundance. We use *evolve* and *evolution* with respect to ecosystems here and elsewhere in this book, not in a Darwinian sense, but in a developmental sense to indicate unidirectional or cyclic ecological change through time. Ecosystem evolution, just like the evolution of species, is sometimes gradual and subtle and at other times rapid or abrupt. A record of the sequential changes in expression that an ecosystem undergoes through time is called its historic ecological *trajectory*. If an ecosystem is impaired, its historic trajectory is interrupted. Ecological restoration allows an ecosystem to resume its historic trajectory. This is similar to a physician assisting in the healing process, so that patients can resume their lives.

During the hiatus caused by *impairment*, the Earth has not stood still. External conditions and boundaries may have changed, and the internal processes of ecosystem recovery may cause ecosystem expression that was not formerly present. Therefore the outcome of ecological restoration is necessarily a contemporary expression and not a return to the past, even though many if not most species may well persist from past to future on most sites. In this way, ecological restoration connects an impaired ecosystem to its future. *We restore historical ecological continuity, not historic ecosystems.* Regardless of how much we try to restore to the past, it never happens. We have no choice in this matter, because we can't control outcomes of restoration without losing the quality of naturalness that we ultimately strive to recover. At best, we can only emulate the past as we restore. The reason for this is that ecosystems consist of living organisms, and life does not run backward. In many restoration projects, the future state emulates the gross structural aspects of the preimpairment ecosystem, but to believe it can ever truly return to that former state—as if time were reversible—is wishful thinking and counterproductive. *We invariably restore ecosystems "to the future."* Consequently, ecological restoration is in some ways a metaphor that should not be taken literally. Nonetheless, it is a powerful metaphor, and a path to follow in troubled times, which has captured the imagination, hearts, and minds of people globally.

All ecological restoration projects are case specific, and it is much easier to restore some impaired ecosystems than others to something that approaches a prior historic state. However, the intent in every case should be to nudge the system back onto its ecological trajectory that—prior to impairment—was in the process of developing toward a sometimes indefinite future. Attempts to restore an ecosystem to its former, historic state are valid and viable as long as it is understood that the outcome will be imperfect, and that in every project our overarching *goals* are to restore historic continuity and ecological wholeness rather than stasis.

We live in a time of increasingly environmental instability. Human-mediated exploitation and abuse of the natural environment, and ongoing changes in climate and other global conditions, dictate that many ecosystems can only be restored to states that contain species substitutions and rearrangements of structure with which we are unfamiliar. Restoring to previously unknown states may seem paradoxical, but this concept is no different from the open-ended nature of ecosystem evolution throughout geological time. What makes ecological restoration distinctive is that we rely insofar as possible on past expressions of the preimpairment ecosystem as our *reference* or starting point and salvage whatever legacies from the past that we can in order to ensure a fully functional, dynamic, and sustainable ecosystem in the future. In particular, we populate the restored ecosystem with species from the predisturbance ecosystem to the extent that contemporary conditions allow. These species coevolved or are otherwise adapted to function seamlessly with one another and the physical environment. They are more likely than other species to assemble themselves into satisfactorily restored and sustainable ecosystems. They provide historic continuity and return an ecosystem to its historic ecological trajectory.

The pace of recent environmental change, coupled with the magnitude of contemporary losses of biodiversity, has quickened ecosystem evolution, generally leading to their impoverishment through simplification and consequent destabilization. Impoverishment is the price we pay for uninformed actions or callous disregard by previous generations and by those of us who indiscriminately transform landscapes, pollute ecosystems, and deplete resources. The vision of restoration, however, gives hope that through our efforts, we can recover ecosystem complexity and once again enjoy the personal, cultural, and economic benefits of functional ecosystems and the biotic grandeur they contain. It suggests that we can undo at least some of the ecological and environmental damage people have caused in the past, and that—despite our ongoing demographic explosion—we can clear new paths for sustainable economic development. The restored ecosystem shown in figure 1.1 demonstrates that ecological restoration is indeed possible and has been accomplished. Such paths must be based on the recognition that our economies depend entirely, in the first and last analysis, on natural capital—the wealth produced by fully functional ecosystems and their biodiversity. Unless we restore this capital, and learn to live on the interest rather than recklessly spending down our reserves, we are doomed to impoverishment at many levels, and possibly to unprecedented economic and ecological catastrophes. The enriching paths of restoration lead to sustainability and reintegration of people with the rest of Nature.

## Some Basic Terms and Concepts

Before going further with our main topic, we pause now to offer explanations of several terms and concepts that recur throughout this book, especially for the benefit of those who are not particularly familiar with *ecology*. The meanings of some of these concepts lack consensus among professionals, and the following discussion explains how we will employ each of these key terms in this book. We have defined additional terms in the glossary that appears at the end of the book, and all terms appearing in the glossary are printed in italics the first time they appear in the text. By and large, we follow usage employed in the first edition of this book (Clewell and Aronson 2007) and in the second edition of van Andel and Aronson's graduate level textbook (2012). A few nuances or changes do, however, occur here, which is only normal given the high speed at which this field is evolving today. We also acknowledge cases where ambiguity remains, and we refer to other published sources for further reading and comparison in these instances.

**FIGURE 1.1.** Wet prairie (foreground) and pond-cypress savanna, restored by D. Borland and A. Clewell for The Nature Conservancy in Mississippi, USA.

## Ecological States

A state is the manifestation or expression of an ecosystem, particularly its biotic *community. Abiotic*—nonliving—aspects of an ecosystem, such as its geology, topography, and so forth, contribute to the state as the setting or backdrop for the biotic community. A system's biotic community is governed by its species composition and *community structure*. The latter is a function of the sizes, *life forms*, abundance, and *spatial* configurations of its species. When we use the word *state* in this book, we refer to an ecological state.

## Process and Function

Organisms and species populations in an ecosystem don't just sit there, as if they were museum specimens. They grow, assimilate, respire, compete for resources, and reproduce. If they are plants, they photosynthesize, absorb water and *nutrients*, and transpire. If they are animals, they move about in search for water and food; they fight or take flight, they compete for mates or advertise for the same. If they are microorganisms, they decompose dead organic matter, release nutrients, fix nitrogen, and transform nitrogenous compounds. All of these activities are biological processes. During the course of any given process, energy is transformed, expended, or stored. Matter is combined, separated into its components, or moved from one place to another. Some of the more important ecological processes include primary production; demographic regulation of species populations by means of herbivory, predation, and parasitism; energy transfers in trophic linkages, nutrient recycling; storage of carbon in humus; soil formation; microclimatic regulation; moisture retention; and many symbiotic interactions, such as animal-mediated pollination and seed dispersal, *mycorrhizal* exchanges of nutrients and energy, and nitrogen fixation by microorganisms living in symbiotic relationships with more complex organisms. These are processes that we associate with ecosystems. We associate other ecological processes with the biosphere level of organization, such as the generation of atmospheric oxygen during photosynthesis, which of course can be studied at the level of ecosystems. Another ecological process of significance to the biosphere is thermal regulation of the Earth's atmosphere, mainly from transpiration, which reradiates heat into space. When a biological system at any level of organization (cell, organism, community, biosphere) participates in a biological process, we say that it is functioning or functional, which indicates that biological *work* is being performed in the sense that physicists use the term.

In recent years, some ecologists and environmental economists, lawyers, and other professionals concerned with natural resources have used the terms *function* and *ecosystem function* in a distinctly different manner from ours to designate collectively those natural ecosystem services that benefit people and their socioeconomic well-being. Examples of ecosystem services are retention of potential flood waters, improvement of water quality, erosion control, provision of range for grazing by domestic livestock, provision of *habitat* for desirable wildlife, and venues for recreation. To avoid confusion, we use the term *ecosystem services* to designate socioeconomic benefits that people derive from ecosystems, and we avoid using the term ecosystem function altogether in this book. This practice conforms to that adopted by the Millennium Ecosystem Assessment (MA 2005) and by the more recent United Nations initiative, The Economics of Ecosystems and Biodiversity (TEEB 2010). We prefer the word *service* in this context, because it connotes that there is a provider and a beneficiary, whereas the word function lacks that inference. The word and the notion of ecosystem function is widely entrenched in discourses pertaining to ecosystem services, their benefits as perceived or enjoyed by people, and the values assigned to ecosystem services by economists and social scientists. Interested readers are warmly referred to the TEEB study for a more thorough consideration of the topic, and especially to de Groot et al. (2010), which is the first chapter in the TEEB study's foundational report.

*Ecosystems*

The basic unit of ecology, and thus of ecological restoration, is an *ecosystem*. An ecosystem is a prescribed unit of the biosphere that consists of populations of living organisms that interact with each other and with the physical environment that sustains them. A. G. Tansley (1935), who first coined the term, described an ecosystem as "the whole system, including not only the organism complex, but also the whole complex of physical factors forming what we call the environment." The living organisms in an ecosystem—plants, animals, and microbial forms of life—collectively comprise its *biota*. The abiotic (nonliving) infrastructure of an ecosystem, consists of the physical environment, such as the substrate or soil, nutrients, water in all forms and its ionic salt content, and of energetic processes and their results, such as hydraulic movements, climatic expressions, and fire regimes. The distinction between the biotic and abiotic components of an ecosystem is more pedantic than real, because organisms and inert materials are constantly interchanging or altering each other as they participate in tight feedback loops. For example, soil formation is governed by numerous interactions between living organisms, dead organic matter, water, atmospheric gasses, and the mineral substrate. Likewise, the *microclimate* associated with an ecosystem is the product of biota influencing the regional climate on account of transpiration, shade, and wind reduction.

Ecosystems are generally circumscribed to display a measure of internal consistency with regard to their species composition, community structure, and abiotic features. The biota is usually subdivided into recognizable biotic communities, such as the plant community, the soil microorganism community, or the *zooplankton* community. Species with shared traits within a given community are designated as *functional groups* or guilds.

Ecosystem is sometimes used in a collective sense to designate a particular kind of ecosystem that occurs repeatedly in an ecoregion, such as *riparian* forest, alpine tundra, or tidal marsh. A more appropriate term for this concept is simply *ecosystem type*. A related term is *biome*, which is applied to a large area in which a particular ecosystem type prevails, such as the southeastern pine *savanna* biome (USA) and the Atlantic forest biome along the eastern coast of Brazil.

Ecosystems are complex, and even within a given biome or ecosystem type, no two of the same kind are ever alike, at least at finer spatial scales. This complexity arises from heterogeneity in the physical environment, *stochastic* variations in ecological processes, and the differential effects of *stresses* and disturbances on an ecosystem from natural and *anthropogenic* causes.

Ecosystems are open, not closed systems. They do not exist apart, as if they were isolated islands. They interact with each other, and most are not readily distinguished or delineated from each other. Seeds, spores, and mobile organisms move regularly among ecosystems. Water flows through one ecosystem into another, carrying dissolved nutrients, mineral sediments, detritus, and whole organisms, leading to fluctuating patterns of availability and quality of resources. Impairment in one system can impact surrounding ecosystems. Boundaries of an ecosystem are designated for convenience by an ecologist or another professional to demarcate an area of interest or study. Its circumscription is necessarily subjective, because the biosphere is an interconnected and interacting whole that defies partition, but this objection does not detract from its usefulness. An ecosystem can be any size, although it is usually much smaller than that of a biome and usually consists of a discrete location within a catchment or a readily identifiable zone or stratum within a water body. One reason to circumscribe an ecosystem is to designate boundaries for an ecological restoration project.

Two interlinked descriptors of ecosystems are *integrity* and health. Integrity is the state of an ecosystem that displays characteristic biodiversity in its species composition and community structure and sustains normal ecological functioning (SER 2004). *Ecosystem health* was described in the *SER Primer* (SER 2004) as the "condition of an ecosystem in which its dynamic attributes are expressed within 'normal' ranges of activity relative to its ecological stage of development." Both terms are qualitative generalizations that resist empirical verification. Cross-disciplinary teams of ecologists and human health researchers considered that ecosystem health can be evaluated in terms of system organization, *resilience*, and vigor, and all of these are characterized by an absence of signs of ecosystem distress (Costanza 1992; Rapport et al. 1998). We refer to ecosystems as being *intact* or whole if they display integrity and health. Conversely, *degradation*, damage, destruction, and transformation all represent deviations from the normal or desired state of an intact ecosystem.

## Production Systems

A *production* ecosystem, as opposed to a natural ecosystem, is a unit of land or water that is transformed—often simplified in ecological terms—and then managed by people to produce crops as commodities with market value or for direct consumption and *subsistence*. In the process, the site or system is normally manipulated in an agronomic, aquacultural, or engineering sense and receives subsidies of energy and materials. Sources of energy may include the work of domestic animals and the use of fossil fuels for operating equipment. Material subsidies may include lime, manure, and compost, or applications of synthesized agrichemicals such as mineral fertilizers, pesticides, and herbicides. Examples of production ecosystems (or *production systems* for short) include agricultural lands

dedicated to row crops, vineyards and orchards, tree plantations, biofuel plantations, impoundments for the production of fish and other seafood by aquaculture,
intentionally managed meadows and pastures for grazing by domestic livestock,
and food plots that are prepared and sown in game reserves. Nonnative species
are commonly introduced into production systems for their commercial values.
Many *agroforests* are production systems for that reason. Landscapes that are dominated by agricultural production systems can be called *agriscapes* and comprise
one category of semicultural landscapes, the topic of chapter 6.

Relative to the natural ecosystems that formerly occupied a site, production
systems are characterized by net reductions in species composition, community
structure, and the capacity to provide a broad array of ecosystem services for people. These advantages are sacrificed for the production of one or a few commodities (e.g., timber) or, nowadays, to provide *ecosystem services* (e.g., long-term carbon storage). The management of production systems sometimes causes adverse
environmental impacts, such as the contamination of water bodies from the runoff of agrichemicals or discharge from animal feedlots. At the site level, they also
tend to be low in biodiversity and provide little habitat for wild plants or animals.

Production systems are neither self-organizing nor self-sustaining. To maintain their productivity, they require management, such as periodic harrowing or
plowing, competitive weed control, mowing, thinning, fertilizing, predator exclusion, and chemical pest control. They may rely on civil engineering, such as the
excavation of canals or ditches; the installation of drainage tiles, weirs, culverts,
pipes, and pumps; and the construction of levees or dikes. Riprap, gabions, and
other permanent engineering features may be needed to stabilize substrates. All
of these so-called improvements require periodic operation, maintenance, and
eventual replacement.

## Landscape

We use the term *landscape* in this book to designate two or more ecosystems that
interact with each other and display a measure of ecological cohesiveness in a
given location, such as a river catchment (or watershed); a portion of a mountain
range; a stretch of coast including dunes, tidal marshes, and lagoons; or any other
geomorphological unit, regardless of size. Landscapes consist of assemblages of
ecosystems and smaller landscape units "that produce patterns that are repeated
and recognizable in space" (Forman and Gordon 1986, 11). This landscape concept, which was adopted in the *SER Primer* (SER 2004), embraces both ecosystems and the frontier or transition zones between them, called *ecotones*. We
use the term landscape to include seascapes, riverscapes, or other "scapes" as determined mainly by *geomorphology*. There are of course many other, more sub-

jective, definitions of the word landscape from geography, art history, and other fields. For a fuller discussion, see Aronson and Le Floc'h (1996a, 1996b).

## Landscape Restoration

Some or all ecosystems that comprise a landscape can be subjected to restoration treatment simultaneously or sequentially with the resulting intent of landscape-scale restoration. However, ecosystems remain the basic unit and focus of ecological restoration. The administratively coordinated restoration of more than one ecosystem in a landscape is called a restoration program, whereas the ecological restoration of an individual ecosystem, whether or not it is part of a landscape restoration program, is called a restoration project. The intent of landscape restoration is to restore a mosaic of interacting ecosystems in order to recover natural and, in many cases, cultural values that are not realized, at least in their entirety, from the restoration of a single ecosystem, and also to recover flows of ecosystem services from multiple ecosystems. This process has been called the reintegration of fragmented landscapes (SER 2004; cf. Saunders et al. 1993). To address these complex issues, we must consider not only natural but also semicultural ecosystems (chap. 6). Under these circumstances, the landscape mosaic is treated as if it were a single, albeit heterogeneous, ecosystem for restoration purposes. Landscape-scale restoration receives more attention than separate restoration projects that only focus on a single ecosystem. The reintegration of disconnected and fragmented landscapes, as with wildlife corridors and free-flowing salmon streams, is a powerful impetus for initiating a landscape restoration program. Some large conservation organizations, including the International Union for the Conservation of Nature (IUCN) and the World Wildlife Fund (WWF), promote landscape restoration under the aegis of what they call the "ecosystem perspective" and "forest landscape restoration" (Rietbergen-McCracken et al. 2008; GPFLR 2012).

## Ecological Reference

An ecological *reference* indicates the intended characteristics of an ecosystem after it has undergone ecological restoration. The reference may consist of one or more intact ecosystems or "reference sites" or their ecological descriptions. If these are unavailable, a reasonably satisfactory reference may be assembled from indirect evidence, as discussed in chapter 7. References sites, their ecological descriptions, and/or indirect evidence contribute to the preparation of a *reference model*, which informs the development of restoration project plans and also serves as a benchmark, a source of inspiration, and a tool for consensus building.

## Sustainability

From an ecological viewpoint, sustainability refers to the persistence of a self-sustaining ecosystem that has sufficient resilience to recover to an intact state, should it suffer from disturbance. In a socioeconomic context, sustainability is the application of sound ecological principles in order to derive ecosystem services on a continuing basis without causing harm to ecosystems that provide these services.

## Practitioners and Professionals

In this book we call people practitioners if they perform biophysical *interventions* at project sites to assist the recovery of impaired ecosystems. Project sponsors and professionals other than practitioners may be intimately associated with restoration practice, including administrators, project planners, project managers, plant nursery personnel and other providers of biotic stocks, social scientists concerned with *stakeholder* issues, and natural scientists who conduct preproject inventories and postproject *monitoring*, and public officials with project oversight responsibilities. These professionals are sometimes identified as practitioners in a broader sense of that term than we use in this book. We use *professionals* here and elsewhere in this book in the sense of persons with competence, whether or not they are formally credentialed or materially compensated for their work.

# Chapter 2

# Values and Ecological Restoration

Why do we restore ecosystems? Why are people attracted to restoration as a career or as a principal focus of their professional work? Restoration is risky, complex, and frustrating long-term work that requires patience and dedication. Working conditions at project sites can be challenging. Other professions offer steadier work and better pay. It can even be difficult for restorationists to explain what they do when others ask because the phrase *ecological restoration* has entered mainstream discourse as a feel-good buzzword without girding from professional standards. And even when restorationists are able to explain what they do quite clearly, they may be hard put to express succinctly why exactly they do it and what it means to them.

Behind the discomfort in addressing these questions lie our values. People choose to become restorationists for a variety of reasons. Most would say it is because of the urgency of the threat to ecosystems, the environment, and the planet. They want to be part of the solution and not contribute any more than absolutely necessary to the causes of ongoing environmental *degradation* and biodiversity loss. Many would add that assisting ecological recovery fulfills other deep-seated values, satisfies diverse aspirations, and gives meaning to their lives. Environmentally concerned individuals who are not restoration practitioners themselves may be keenly interested in promoting ecological restoration projects that improve ecosystem services, promote environmental education, and provide recreational opportunities; or that enhance the aesthetics of natural areas, recover revered or sacred sites, and help redeem the interconnectedness of the biosphere. These are only a few of the values that ecological restoration satisfies. Some values, as we shall see, are fulfilled directly by the performance of ecological restoration. Most values are satisfied later by ecosystems after they are restored.

Some values are subjective and emotional, like improving aesthetics, while others are objective and pragmatic, like improving flows of ecosystem services. Some values satisfy individuals, like the contentment one experiences from expending effort to reverse environmental damage, while others are collective such as the social cohesiveness that develops among volunteers who work together on a community-based restoration project. In order to consider the many values accruing from ecological restoration in a holistic yet organized manner, we categorize them in a *four-quadrant model for ecological restoration* in figure 2.1. This model was adapted from a schematic diagram devised by contemporary philosopher Ken Wilber (2001), which recognizes categories as either objective or subjective and individual or collective. Wilber's generic model is applicable to numerous disciplines and, for example, has been used effectively to portray the new discipline of integral ecology, which serves as a framework to allow ecological issues to be viewed from multiple perspectives (Esbjörn-Hargens and Zimmerman 2009).

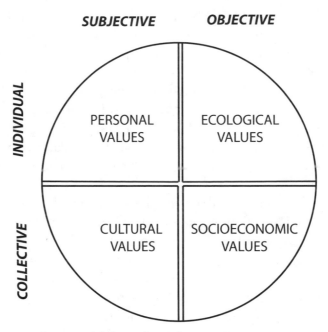

**Figure 2.1.** Four-quadrant model for ecological restoration.

As viewed from left to right, the model consists of two hemispheres, one pertaining to subjective values and the other to objective values. Objective values can be measured and analyzed empirically, whereas subjective values express opinions and emotional responses that resist direct empirical measurement, even though economists and other social scientists attempt such measurement with some success. As viewed from top to bottom, the model consists of an upper hemisphere that is relevant to individuals and a lower one that pertains to people collectively in social or cultural groups.

The upper left quadrant represents our emotional reaction to ecological impairment. The latter may incur outrage that our culture would allow impairment to happen, or impairment may insult our aesthetic sensibilities. Our responses to such reaction may lead us to undertake an ecological restoration project directly as practitioners or indirectly as project sponsors, financiers, or citizens who express political opinions that influence public policy. All such responses provide the satisfaction of knowing that we did something proactive to resolve a problem that concerns us all.

The upper-right quadrant represents ecological features that we value as attributes of intact, healthy ecosystems and ecological landscapes; and to the biosphere, and its integrity or interconnectedness. This quadrant expresses our rational response as individuals to ecological impairment based on our understanding of natural areas, their biophysical components, and ecological processes. Ecological restoration allows the redemption of these lost attributes and the values we assign to them.

The lower right quadrant represents our collective socioeconomic values in regard to ecosystem services that were curtailed or lost from ecosystem impairment. On account of impairment, we may have placed ourselves at greater risk to flooding, or we may have to pay higher utility charges to purify water. Seafood we prefer may have become scarce and its purchase price more costly. In other words, we have collectively suffered losses in our socioeconomic values on account of ecosystem impairment that, in turn, reduce our standard of living and well-being and, at times, may threaten our survival. Ecological restoration allows us to recover these socioeconomic values in terms of increased flows of ecosystem services. These services promote a stronger economy, reduce economic distress, and increase the capacity for social integration.

The lower left quadrant represents our collective cultural values that were impinged by ecosystem impairment. We feel deprived by the loss of an iconic site that was damaged, such as a public park or sacred site. If we respond as a community to restore what we have lost, our dedication as we work for a common purpose will strengthen our bonds as neighbors and our social cohesiveness. We will learn more about our environment. Children who participate in restoration activities will absorb important lessons in natural history and ecological literacy (or ecoliteracy). Their appreciation of biodiversity gained from school books or the media deepens with that gained from direct experience. Development of an environmental ethic is only a small step away, once the profundity of the biodiversity concept is acquired by those working on a restoration project. All of these are values we collectively satisfy as we participate in ecological restoration, directly or indirectly. Satisfaction of these values strengthens the nexus between Nature and culture.

Many but not all subject values are fulfilled by the performance of ecological restoration, whereas all objective values are satisfied by ecosystems after they have been restored. With the exception of some aesthetic values, nearly all personal values are fulfilled by performance—the process of ecological restoration. Some cultural values are fulfilled by performance, particularly the development of social cohesiveness and those values pertaining to ecoliteracy that are satisfied from hands-on activities at project sites.

Figure 2.1 is drawn with double lines separating the four quadrants, indicating a degree of separation and independence. This separation is intentional and represents the analytical inclination of professionals to focus on the one quadrant that best represents their professional interest. Wilber (2001) cautioned that those who ignore the whole and concentrate on a single quadrant are unable to see how the pieces fit together. They wear self-imposed blinders. He calls them *flatlanders*, referring to pre-Renaissance intellectuals who ignored mounting evidence that the Earth was round. Instead, planet Earth, and ecological restoration, must be viewed and conceived in a holistic manner.

Everyone involved in a restoration project naturally brings a personal perspective and with it a particular set of values. For example, ecologists and conservationists are likely to be attracted to the upper right-hand quadrant of ecological values. For them, the intent of restoration is ecological recovery. In contrast, personnel who are responsible for management of natural resources are more interested in replenishing ecosystem services, such as providing clean water or habitat for imperiled species. They are not necessarily concerned with the ecological details and instead are drawn to the quadrant of socioeconomic values. Other people are motivated by personal reasons and gravitate to the upper left-hand quadrant. These people engage in restoration project work because it satisfies their inner personal needs to reconnect with Nature or atone for the environmental ravages that were perpetrated by their culture. Yet others are culturally motivated, such as the teacher who is thrilled to know that a restoration project in a schoolyard is raising ecological literacy to a degree that could never be attained in the classroom.

Regardless of one's personal perspective, the essential point is for everyone involved in a restoration project to be aware of the values from all four quadrants. This broad-focused awareness makes more meaningful a given project that may have been intended initially to fulfill values from a single quadrant. The stated goals of many restoration projects pertain to the satisfaction of values for only one or two quadrants. We contend that every well-conceived, holistic, ecological restoration project satisfies values from each of the four quadrants in figure 2.1, regardless of its stated goals. We now discuss some of the more important values that are satisfied by ecological restoration in each of the four quadrants.

## Ecological Values

The crucial ecological values of ecological restoration from an objective, scientific point of view pertain to the recovery of impaired ecosystems to wholeness in terms of their ecological integrity and health. These values, in turn, are attained by interventions performed at a project site to recover an ecosystem's biophysical conditions that were impaired. Thereafter, other attributes begin to emerge without assistance from practitioners on account of biological activities common to every ecosystem. They include the restarting of ecological processes; the development of ecological complexity, self-organization, resilience, and self-sustainability; and reestablishment of historical continuity with respect to ecological trajectory. These attributes are prized by ecologists as important ecological values and will be described in greater detail in chapters 4 and 5. Ultimately, these values contribute to the support of the entire biosphere, principally in terms of regulating atmospheric oxygen and carbon dioxide, facilitating thermal reflectance of solar radiation, and providing habitat for imperiled species. Lovelock (1991) introduced the allegory of the ancient Greek goddess *Gaia* to emphasize the interconnectedness of the biosphere, suggesting that the Earth itself is a self-regulating organism. The biophysical reality of Gaia can be easily challenged by materialistic science but not its usefulness as a reminder of the importance of holistic approaches to our biosphere. We assert here—and throughout this book—that ecological restoration contributes significantly to biosphere support.

Ecosystem recovery may seem to be an unquestionably obvious goal to many readers. But we need to be scrupulous here and acknowledge that, like all goals, those of restoration are informed by a particular set of values. Davis and Slobodkin (2004a, 2004b) and Lackey (2004) assert that terms such as damage, repair, integrity, and health are all subjective, value-laden terms lacking in scientific objectivity. Lackey wrote,

> For example, one person's "damaged" ecosystem is another person's "improved" ecosystem. A "healthy" ecosystem can either be a malaria-infested swamp or the same land converted to an intensively managed cornfield. Neither condition can be considered "healthy" except through the lens of an individual's values and preferences. (2004, 45)

We agree, unless the terms and conditions of ecological restoration are qualified.

In chapter 1, we identified what we meant by ecosystem integrity and health in a scientific context with regard to ecological impairment and recovery. Of course these terms can never be entirely value free, nor should they be. They are more fully qualified, however, in measurable ways in chapters 3 and 5.

## Personal Values

To introduce personal values, we pose the question to the reader: Why would you, on a profoundly individual and personal level, want to restore ecosystems? Three answers that we have often heard are summarized in this section. Each answer reflects a somewhat different value.

*"Because restoration gives me a meaningful way to reconnect with Nature."*

This question addresses the reentry value, a term we borrow from Bill Jordan, who called restoration a vehicle for establishing reconnection with—or reentry into—Nature. Jordan (1986) explained that there are many ways to come into contact with Nature. Contact can be as a visitor in terms of hiking, backpacking, canoeing, mountain climbing, bird watching, or Nature study. Contact can be more exploitive or interactive, involving hunting, fishing, agriculture, gardening, or gathering wild plants for food or dye making. Making contact with Nature is not as profound or satisfying as reentering Nature. Jordan (1986, 2) wrote that none of these ways of contact with Nature "offers complete immersion in nature through the exercise of the full range of our abilities as human beings," that is, as makers, inhabitants, and bona fide members of the natural world. Restoration allows us that opportunity. In this regard, restoration affords the practitioner the opportunity to participate in natural processes from the inside out rather than as a technician who is called in from afar to fix a problem. Practitioners share a bond of kinship with their landscape. Reentry or reconnection lifts the practitioner beyond the despair that is understandably pervasive among preservationists because restoration aims to actively repair environmental damage. This effort imbues the practitioner with the satisfaction of doing something positive that reverses environmental destruction.

In affluent, industrial regions of the world, the urge to reenter Nature is deep-seated and is generally equated with recreation. We *re-create* ourselves by engaging in pleasurable activities that remove us temporarily from the stress or humdrum of making a living and the turmoil of daily life. There is another undeniable aspect of reentry that is not as well appreciated. It is the aesthetic appreciation of Nature. It is not just a matter of absorbing an occasional view of a majestic mountain range but also the simpler aesthetic response to the sleek symmetry and flashing colors of a fish as it swims by—or as it is reeled in by a fisherman. Consider the wonder expressed as a child pokes at a ladybug: that too is aesthetic in the deepest sense. Ecological restoration immerses the practitioner into a world of aesthetic feelings and impressions that are revealed bit by bit as an ecosystem emerges from impairment and returns to health. One's aesthetic sensibilities—and, in the best instances, cultural longings and identity—can draw us into performing ecological restoration. We will return to cultural values later.

*"Because we have an environmental crisis on our hands, and I'm going to do something about it!"*

This may be the main personal reason why people become restoration practitioners. It is what writer/journalist Paddy Woodworth (personal communication, 2007) called the *environmental crisis response value* (or enviro-crisis response, for short). This response is recognition that the world and its human population are in trouble environmentally and that we had better act now to ensure our future well-being. Ecological restoration is an especially attractive solution for those who take personal responsibility for the environmental harm their culture perpetrates, because it reverses environmental ruination in an easily observable manner (Jordan 1994; Higgs 1997).

*"Because it allows me to experience oneness with Nature."*

Practitioners may come to realize that they are intimately meshed with, and ultimately indistinguishable from, the ecosystem on which they are working (Clewell 2001). This is a powerful intuition or unreasoned awareness that may occur unexpectedly in the midst of routine project work. It is a very personal, subjective experience that cannot be rationalized and is difficult to describe. We do not know how common or rare this experience is; however, we suspect it happens more than occasionally and that practitioners find it too personal to relate to others. When we take field trips in conjunction with restoration workshops, where participants spend an afternoon *outplanting* nursery stock at a restoration site, we regularly ask participants if they feel a spiritual connection or oneness with the ecosystem that they are restoring. Invariably, some confirm that they do. We allude to this awareness as personal transcendence to a profound level of consciousness or oneness with Nature.

Reentry, environmental crisis response, and the experience of oneness are all powerful incentives that reach deep into our psyches. We may rationalize that we entered the field of restoration because we took a course in that subject in college or because we were offered a job with a firm or institution that was doing restoration. The chances are that we took that course or considered that job because it resonated with our yearnings to fulfill of one or more of these related values.

Is it really necessary for someone to acknowledge these motivations before becoming a practitioner? Perhaps not, but we feel that anyone who enters a new discipline such as restoration should be consciously aware of his or her motiva-

tions and recognize the values that made restoration attractive. It is not only practitioners who benefit from this deeper understanding. It also helps anyone who
is engaged in any way with restoration project work, including administrators,
financiers, and officials who make public policy regarding natural resources. Everyone can gain from this same grounding in values if they are to appreciate the
importance of restoring ecosystems and retain the patience and devotion needed
to see it through.

## Socioeconomic Values

We depend on and value Nature for the air we breathe, the water we drink, the
food we consume, and the raw materials that supply our domestic well-being and
our urban industrial bases. We also depend on and value Nature to stabilize the
soil and keep it from eroding, to detain surface runoff that would otherwise cause
flooding, and to provide many other ecosystem services. The degree to which we
are aware of these values from Nature depends on where we live and what we
do. Many urban dwellers are only vaguely familiar with them. Rural residents,
particularly those from impoverished areas, are likely to be much more knowledgeable because they are directly dependent on *natural goods and services* for
their survival and well-being. Furthermore, as will be discussed in chapter 10, the
true value of Nature is not fully recognized in current market transactions and by
public policy—far from it. However, given the growing awareness and concern
about environmental problems, including the accelerating and irreversible loss of
biodiversity and global climate change, it is clear that many people are reevaluat-
ing personal and societal values in this arena.

Ecosystems provide a wide array of natural products and services that are useful to people and on which all economics depend. These goods and services are
free to people and available without costs of production. In contrast, goods such
as agricultural products and manufactured items must be purchased. Engineering services must also be purchased. The array and importance of natural goods
generated by natural ecosystems are becoming better known on account of recent
interdisciplinary works such as those of Westman (1977) and Daily (1997) and the
monumental Millennium Ecosystem Assessment project (MA 2005) and, more
recently, the TEEB study (The Economics of Ecosystems and Biodiversity: TEEB
Foundations 2010; TEEB 2011; http://www.teebweb.org/), which addresses a
broad audience, including economists, administrators, finance ministers, policy
makers in general, the general public, and business leaders. The more important
and obvious natural goods are listed in table 2.1. An important natural good in
warmer regions is bamboo, shown harvested in figure 2.2 and awaiting manufacture into newsprint.

TABLE 2.1.

*A sampler of ecosystem goods*

Wood for construction purposes such as lumber, poles, and cross ties

Thatch for roofing materials and matting, consisting of palm leaves and grasses

Fiber for textiles and rope

Firewood for domestic cooking and heating and for charcoal production

Forage for grazing by domestic livestock on meadows and other rangelands

Fodder and silage for domestic livestock

Medicinal plants and pharmaceuticals

Dyestuffs for use in textiles, foods, and cosmetics

Exudates such as gums, resins, and latex

Honey and oils

Vegetable foodstuffs such as roots, nuts, berries, and mushrooms

Natural products for use in rituals, such as incense

Seafood including finfish, shellfish, macrophytic algae, and marine mammals

Bushmeat and animal products from a wide range of vertebrate and some invertebrate animals consumed for food and used for hides, sinews, bones, tusks, blubber, and other materials

**FIGURE 2.2.** Bamboo harvested from natural ecosystems in India and stacked in preparation for the manufacture of paper.

It is true that synthetic or cultivated replacements exist for most of these natural goods, at least for affluent people who benefit from economic globalization. But for much of humanity, few or no substitutes are available locally. Furthermore, substitution incurs global costs, demonstrable by cost-benefit analysis, and should be avoided where possible. This reflection is even more pertinent if we also consider the gamut of natural ecosystem services provided by ecosystems. Ecosystem services of value to people are listed in table 2.2.

Natural goods and natural services are collectively called *ecosystem services* by those who manage and regulate natural resources. Our economic structure and our social structure both suffer when ecosystem services are threatened. In chapter 10 we will explore the use of ecological restoration to augment the array and magnitude of natural goods and services as a means to alleviate poverty, joblessness, and the growing problems of ecological refugees caused by local environmental degradation and scarcities of ecosystem goods and services.

## Cultural Values

We restore ecosystems to satisfy values that are shared collectively within a culture. For example, much restoration is dedicated to the recovery of impaired ecosystems in iconic places such as parks and preserves, where people gather to enjoy Nature-oriented recreation and leisure, or in sacred places that have spiritual or religious significance (Ramakrishnan 1994; Desai 2003). Figure 2.3 shows a sacred grove on a hilltop in India, and figure 2.4 shows villagers who visited their sacred grove for spiritual purposes. We may participate in local environmental stewardship programs where we repair public lands that have been impaired by excessive visitation or intense recreational use, as happens on ski slopes or trails and dunes traversed by all-terrain vehicles. We may join a community effort to clean up and refurbish a soiled stream channel, and restore a riparian forest ecosystem along its course, in an effort to entice the return of native salmon to their breeding areas.

Participation in community-based restoration develops a strong sense of place and a sense of community that occur when like-minded people join in a restoration project to improve their local landscape (Clewell 1995). Projects of this sort have led to civic celebrations and even to a wedding of practitioners (Holland 1994). A place-based movement that was fueled by community-based restoration efforts was under way in the United States in the early 1990s as a response to the post-World War II habit among Americans to change residences and geographic locations frequently. This migratory lifestyle cast people into landscapes where they had no prior sense of attachment and little understanding of the natural features that these landscapes offered for their support. Environmental quality

TABLE 2.2.

*A sampler of ecosystem services*

Protection of water recharge areas by vegetation that absorbs rainwater and snowmelt and that detains surface runoff and allows it to percolate into aquifers, from which water can later be extracted for use

Detention of potential floodwaters in wetlands and as groundwater storage

Reduction of soil erosion and consequent reduction of eroded sediments through the soil-binding capacities of roots and the soil crusts formed by microorganisms

Transformation of excess nutrients, including denitrification and the storage of mineral nutrients in biomass and detritus

Immobilization of contaminants, such as heavy metals, agrochemicals, disease-causing organisms, and pollutants in stormwater and industrial discharge, by organic matter and other colloidal materials in soil to which these contaminants are adsorbed

Cleansing of particulates from the air by the filtering action of forests and other terrestrial vegetation

Cleansing of particulates in water by adsorption to organic surfaces and by the stilling of turbid water by aquatic and emergent vegetation and consequent settling of suspended solids

Reduction in noise pollution by the baffling effects of mainly arboreal vegetation

Renewal of topsoil through the incorporation of humus into mineral soil or the deposition of peat and muck

Conservation of germplasm (genetic material), such as that of the wild progenitors of cultivated plants and domesticated animals for use in overcoming inbreeding and for introductions of genes to induce disease resistance and to develop new economic varieties; also the conservation of alleles (genetic stocks) as an aspect of biodiversity

Provision of habitat for pollen vectors, particularly of domesticated crops, which commonly need natural habitat for completion of their life cycles

Provision of habitat for predaceous arthropods (e.g., insects, spiders) or other predators of crop pests, which also need natural habitat for completion of their life cycles

Provision of habitat for valued wildlife, including rare, endangered, threatened, and red-listed species, as well as game animals, including fish, in areas where hunting and fishing are practiced

Buffering of acidity in soil and water

Regulation of the quantities of atmospheric gases, including oxygen and carbon dioxide

Provision of recreation areas, including for ecotourism

Offsetting or dampening extremes of climate by the dissipation of solar radiation as heat

Buffering coastlines and shores from wave action, tsunamis, and storm surges

deteriorated as new residents were insufficiently grounded to make wise decisions on land use. This same approach is being used by environmental *nongovernment organizations* (NGOs) in tropical villages in South America, Madagascar, and

**FIGURE 2.3.** Sacred grove on hilltop in west-central India, surrounded by pasture near the entrance to a tribal village. This grove is undergoing ecological restoration for its expansion and to recover ecosystem health that was lost due to unrestricted grazing by domestic cattle.

**FIGURE 2.4.** Tribal villagers in India leave a sacred grove after performing *pujas* (acts of reverence) for local gods who reside there.

elsewhere, in regions at the edge of primary forest where most people are new immigrants—ecological refugees—with few or no cultural roots or sustainable agricultural traditions.

Education and investigation in the pursuit of knowledge are deeply ingrained cultural values. Much restoration has been conducted in schoolyards and college campuses for the purpose of raising ecological knowledge (Orr 1994). The celebrated Curtis Prairie at the University of Wisconsin Arboretum was restored in the 1930s to allow access to a *prairie* for study by university students in ecology at a time when transportation was limited (Jordan 2003).

A major reason for restoring ecosystems is to provide places where biodiversity can be protected—also known as "restoring habitat." Biodiversity has broad cultural appeal, which is evidenced by numerous school and television programs that highlight biodiversity. The phrase "biodiversity value" enjoys considerable use in both popular and scientific media. Participants in restoration projects receive an inside view of biodiversity that cannot be appreciated from watching Nature shows on cable television.

Ecological research can be viewed as a cultural endeavor in the same manner as is education. The contentions by Harper (1987) and Bradshaw (1987) that ecological restoration can be an acid test of ecological theory are frequently cited. This opinion is optimistic because the restoration process is sequential and cumulative, which limits the ability to test single variables. However, research designs and statistical methods can help restorationists deal with these problems (Osenberg et al. 2006).

The values that motivate restoration, then, are multiple and diverse. People in most cultures can relate to at least some of them. This is grounds for optimism that ecological restoration, once effectively communicated to the general public, will have a smooth passage to broad global acceptance as having a major role in coping with the daunting environmental challenges of the twenty-first century.

## Holons and Realms of Organization

The four quadrants of figure 2.1 are redrawn in figure 2.5, to which is added two additional features borrowed from Ken Wilber (2001). One consists of four axes, one for each quadrant, shown as dashed lines that extend outward from a central hub. Selected values are shown along each axis. These values are arranged so that each builds upon—or incorporates—its neighboring value that lies closer to the hub. Wilber calls each axis a holon. The center of the four-quadrant model is the hub where holons originate. This hub is the vision we have when we contemplate a new ecological restoration project, and it is has a quality of a Platonic ideal.

The principle of holons, whereby each successive element incorporates all pre-

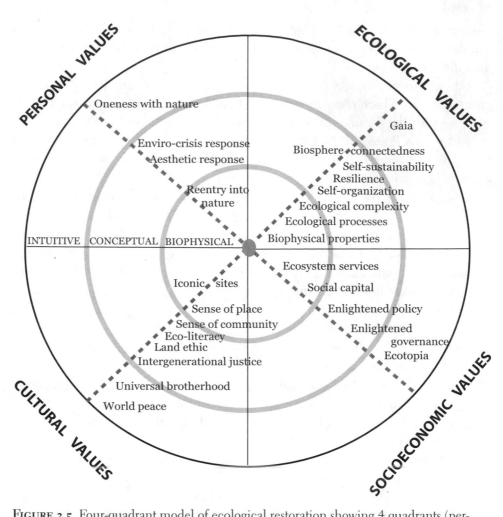

**FIGURE 2.5.** Four-quadrant model of ecological restoration showing 4 quadrants (personal, ecological, cultural, and socioeconomic values), 4 holons (dashed lines from the hub), and selected values along each holon. Concentric circles indicate divisions between the biophysical, conceptual, and intuitive realms.

ceding elements, is essential to the four-quadrant model of ecological restoration. The elements closest to the hub are necessarily material or tangible and pertain to biophysical aspects in Nature and demographic aspects of human organization. In figure 2.5, these elements are located in what is labeled the biophysical realm. The next elements along holons are conceptual and pertain to the meaning we assign to elements in the biophysical realm. These elements are intellectual and logical in an Aristotelian sense rather than tangible, although they incorporate biophysical elements. Without them there would be nothing to conceptualize. The outermost or intuitive realm reaches beyond ordinary thought processes to insight and comprehension that bypass logic. By the intuitional realm we refer to the bursts of realization that we occasionally experience and that defy explanation. Our conception of the intuitive realm corresponds with the thinking of Thomas Kuhn (1996) who, in his classic book *The Structure of Scientific Revolutions* (122–23), asserted that all science depends on flashes of intuition for the development of hypotheses and the evaluation of data. In this realm, we return to an ideal state that is more akin to what Plato described in *The Republic* (Rouse 1956).

In the quadrant for ecological values, the holon consists of universally occurring ecological attributes, as will be discussed in chapter 5. Such a holon begins with the biophysical properties of an ecosystem (biophysical realm), progresses to ecological processes, and from there advances to attributes of ecosystem complexity, self-organization, resilience, and sustainability, which belong to the conceptual realm. Each of these contributes to biosphere sustainability and interconnectedness or Gaia, which belong in the intuitive realm.

The elements placed along the holon for personal values begin with one's reentry into Nature, which in turn satisfies a personal aesthetic response. These two values help formulate personal responses to environmental crises, which is a reaction to ecosystem *impairment*. As an ecological restoration practitioner, the individual may eventually develop an intuitive sense that he/she is participating in one's own ecosystem and is inseparable from it. This is the realization of oneness with Nature and the understanding of restoring one's own psyche as an integral element of the ecosystem undergoing restoration.

The holon for socioeconomic values begins with ecosystem services in the biophysical realm. The realization of ecosystem services stimulates the development of social capital consisting of people who understand the importance of natural areas that provide ecosystem services. This political base stimulates the development of enlightened public policy and governance to foster natural areas protection, management, and restoration, and that oversees the equitable distribution and wise use of natural resources (conceptual realm). This process leads ultimately to an ideal future nation-state or civilization in which a healthy relationship has been achieved between humanity and the environment. We borrow the title of the novel *Ecotopia* (Callenbach 1975) to designate this ideal state, which is perceived as being imbued with ecological wisdom and populated with a sustainable civilization. Each of these elements could not exist without the previous elements on which they depend.

Elements along the holon for cultural values begin in the biophysical realm with iconic areas, such as public parks and sacred groves with cultural significance. Community-based restoration of iconic areas provokes a sense of place within the community, which, in turn, stimulates (in the conceptual realm) a sense of community. Concomitant with the development of a sense of community is a growing appreciation of biodiversity and increasing sophistication in ecoliteracy. Eventually, a shared land ethic develops along the lines that Aldo Leopold (1949) advocated. These advances generate a realization that sharing of cultural responsibility is essential if cultural environmental values are to be preserved and passed onto future generations. This realization stimulates interest in promoting intergenerational justice. A culture that develops a land ethic and accepts intergenerational justice has outgrown the propensity for greed and has replaced it with an ethic of shared caring and trust as its underlying motivation. Striving for universal brotherhood replaces the urge for individual gain. If shared caring and trust become international, the ideal of world peace becomes a reality without treaties or need for military enforcement. At this point, the holon has entered the intuitive realm.

The four-quadrant model has yet another feature that is not apparent when portrayed in two dimensions in figure 2.5. That feature is the unity that is approached in the intuitive realm. As suggested already, the quests for ecotopia, world peace, oneness with Nature, and interconnectedness of the biosphere are all aspects of a search for wholeness. A true ecotopia could scarcely arise without world peace and without a sustainable, interconnected biosphere—Gaia—to sustain it. World peace is surely related to the inner peace of individuals, which in turn seems related to intuitive experience. In this manner, the four quadrants become one in the outermost realm. Figure 2.5 begins at the base with unity, where the four holons diverge. The pertinent elements for an ecological restoration project are distinguishable, as shown on each holon. Ultimately they are rejoined in the intuitive realm. This unity is indicated when figure 2.5 is redrafted in three dimensions, as shown in figure 2.6.

Shall we then agree that ecological restoration is *not* a one-dimensional exercise? Instead, it is a *holistic* multidimensional endeavor in which elements on each axis or holon of all four of the quadrants we have identified are ultimately, and essentially, *inseparable*, despite our academic and administrative efforts to deconstruct them. Even if unity is not yet achieved, none of the individual elements can be ignored or dismissed. The fundamental idea we elaborate here is that ecological restoration has an extraordinarily positive role to play in multiple arenas of human life. For example, to resolve climate change issues by means of ecological restoration or any other technical strategy will require enlightened public policy, enlightened governance, bordering on ecotopia, and a cultural revolution moving global society toward sustainability and world peace.

We sometimes refer to "the promise of ecological restoration." The phrase may seem poetic and meant only as a feel-good notion, but we foresee it as something real that someday could materialize. The promise of restoration is the potential for culture that embraces ecological restoration with seriousness and enthusiasm to

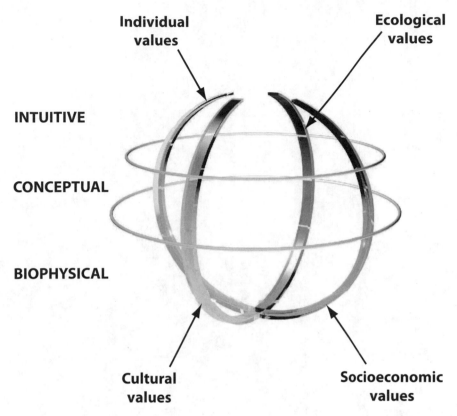

**FIGURE 2.6.** Four-quadrant model of ecological restoration, showing the pending convergence of the four upward-arching holons, which diverged from a central hub at the bottom in the biophysical realm, and which could reunite above at some future point in the intuitive realm.

experience an idealistic existence that is portrayed in the intuitive realm in figure 2.6. Ecological restoration has that much potential, because it is the way we can reciprocate and sustain Nature—as we noted in the opening paragraph of this book—which, in turn, sustains us.

# Chapter 3

# Disturbance and Impairment

We humans are adept at manipulating our biophysical environment. Our ancient ancestors had no choice but to modify their home ecosystems in order to eke out subsistence and secure their survival. In this regard, they were no different from other animal species (see Jones et al. 1994; Lavelle 1997; Rosamund and Anderson 2003; Vermeij 2004). Beavers impound streams and radically alter the environments of catchments. Forest corridors used by Asian elephants are recognizable by characteristic bamboo-dominated ecosystems that develop in response to the mechanical damage caused as these large beasts move about. Termites construct complex edifices with intricate thermodynamic properties and strong impact on soils. Even fungi and bacteria modify their environments proportionately to their small size by exuding enzymes that digest enveloping organic matter.

Today, the burgeoning human population's demand for ecosystem resources worldwide is causing ecological impairment at an increasing rate, and the biosphere and almost all the Earth's ecosystems are degrading in response. The human population has tripled over the past century. Vast numbers of people from poverty-stricken regions remain in subsistence mode, partly as a result of their traditional forms of exploitation of ecosystems in their quest for survival and, even more so, due to flagrant injustice and aggression within and among human societies. Overgrazing by domestic livestock and excessive harvest of wood for fuel are common examples of the first problem. Wars and colonial invasions top the list of drivers for the second problem. Concurrently, a miniscule elite of affluent people, tellingly called "biosphere people" by Indian ecologist Madhav Gadgil (1995), threaten ecosystem integrity, health, and sustainability globally through their excessive demands for ecosystem goods and services from all points of the compass. Ecological restoration attempts to rectify ecological impairment and address the larger problems mentioned just now, and it merits serious consideration as a strategy to help offset and indeed reduce the growing human ecological imprint and *footprint* as we strive for greater social justice and sustainability in this young twenty-first century.

Should restorationists also attempt to repair and recover ecosystems that were harmed by nonhuman-mediated phenomena? Even without "help" from people, mountainsides can fail and bury forests beneath debris. Presumed natural ecological damages, though, are sometimes mediated or intensified by human agency. Mountainsides fail after forests are harvested, exposing soils to unchecked erosion. Mangroves would have ameliorated the ecological damage of the great tsunami in the Indian Ocean in 2004, if these coastal forests had not been previously removed. Some would argue that restoration under these circumstances amounts to environmental meddling and another instance of human artifice and hubris that despoils Nature or ignores natural cycles and rhythms. Aside from such philosophical objections, there may often be reason and economic justification to invest in restoring ecosystems impaired by nonhuman drivers in order to safeguard human welfare, regain the flow of services, and protect a full range of human values. Notwithstanding, nearly all restoration projects address impairment caused by humans, and in this book we concentrate on those situations.

However, it may be more effective, and socially more compelling to seek to restore impaired ecosystems in locations where the threat of pending natural disaster is particularly high, rather than to wait until after it occurs. Property damage in New Orleans caused by hurricane Katrina in 2005 was augmented considerably by the prior destruction of tidal marshes that would otherwise have reduced the impact of storm surges (Costanza et al. 2006). Those who had advocated tidal marsh restoration for many years were vindicated when Katrina destroyed large portions of that city, which has yet to fully recover. It has become clear that near-annual destructive floods in the Mississippi River system will not abate until we restore headwater wetlands in order to retain snowmelt and spring rainfall (Hey and Philippi 1999; Day et al. 2005). Until that time, cities will continue to suffer flooding as occurred in the suburbs of St. Louis, Missouri, in 1993 and in Memphis, Tennessee, in 2011. Those headwaters were drained to expand areas for agriculture over the past two centuries, but the time has come to reappraise those actions.

We continue by dwelling momentarily on the natural history of disturbance. Ecological restoration becomes necessary on account of disturbance in the environment, but not all disturbances require a response from ecological restoration. To the contrary, much disturbance is vitally necessary to maintain ecosystem integrity and to ensure long-term sustainability. Restoration practitioners need to distinguish between the various degrees and kinds of disturbance.

## Disturbance

Disturbance, or perturbation as it is sometimes called, is the technical term used by scientists to describe serious disruptions of ecological functioning and modification of biotic expression, especially at population and community levels. Much disturbance originates from exogenous (external) influences, such as prolonged drought, flooding, or oxygen deprivation; an extreme temperature event, a massive mudslide, fire, or salinity shock. And of course we must acknowledge the damage caused by ill-considered land-use activities, such as excessive stocking of livestock; clear-cut harvesting of multiaged timber; extraction of oil, gas, or minerals; and the installation of urban or transportation infrastructure that ignores the fundamentals of landscape ecology. When caused by anticipated, internal processes—such as canopy gap formation from tree fall in a forest, or exposure of mineral soil by burrowing mammals—disturbances are deemed endogenous. In other words, they contribute to a normal disturbance regime that revitalizes ecological productivity and maintains ecosystem continuity.

Exogenous disturbance events are called "drivers" if they cause net ecological alteration or change, "triggers" if they cause an abrupt change, and "forcing factors" if they cause an ecosystem to switch from a prior stable state to an alternative stable state with a contrasting biota and functionality. Forcing factors are generally anthropogenic and include prolonged intensive grazing by domestic livestock, periodic mowing, fire, and practices that cause modification of the soil, such as compaction resulting from mechanical equipment or concentrations of livestock. Exogenous disturbance factors should not be confused with factors that typify normal environmental conditions for a particular ecosystem. For example, elevated salinity is normal in most marine environments. Excessive evaporation during drought can trigger episodes of hypersalinity, which represent exogenous disturbance events in estuaries and other coastal ecosystems, but not in open oceans.

## Ecosystem Response to Disturbance

Disturbance of many kinds can occur, each with different levels of intensity. They provoke varying levels of damage and elicit a broad array of responses from ecosystems. Disturbance (excluding normal disturbance regimes) can be roughly divided into three ascending categories:

1. *stress* that maintains ecosystem integrity;
2. *moderate disturbance* from which an ecosystem can recover in time without assistance; and
3. *impairment*, a more severe case, where human intervention may be needed

to prevent unacceptable transformation to an alternative and probably less ecologically vigorous state.

Stress temporarily impedes growth and productivity, but it does not threaten the integrity and health of ecosystems. On the contrary, stress may kill or suppress plants and animals that are not native to, or representative of, that ecosystem and that colonized subsequent to the preceding stress event. If unchecked, these atypical or invasive organisms may compete successfully and transform the ecosystem to an *alternative state*. Periodic stress events prevent transformation and maintain ecosystem integrity. Disturbances that recur and serve as barriers to the establishment of uncharacteristic species have been called "stressors" (Lugo 1978), and "disturbance filters" (Grime 2006). Stressors and disturbance filters are counted among drivers or triggers. Stressors exact a toll in terms of productivity and standing biomass, which is the ecological price to be paid for ensuring ecosystem integrity and health. Probably *all* ecosystems are maintained to some degree by stressors.

*Pyrogenic* (fire-generated and regulated) ecosystems are maintained by fire as a stressor, which normally prevents woody plants—or at least those that attain large size or wide cover—from becoming established. The aerial portions of plants are consumed by fire but recover new growth directly from their perennating structures (*rhizomes*, etc.). Prolonged fire cessation generally causes transformation of a pyrogenic ecosystem to an alternative and often highly contrasting state. Prescribed fire is a commonly used management technique on both private and public lands, including some national parks in the United States and Australia (Fernandes and Botelho 2003; Nesmith et al. 2011; Taylor and Scholl 2012).

Other kinds of stressors include freezing temperatures; salinity shock in tidal environments where the saline content of the water is subject to rapid changes in concentration; and anoxia (absence of oxygen) in the soil, which commonly occurs seasonally in wetlands. In each of these situations, stress-intolerant colonizers are killed or controlled. Stress events generally recur with anticipated regularity and comprise what is called a disturbance regime. These recurrent disturbances are characteristic of an ecosystem and can therefore be considered as long-standing, evolutionarily stable precedents that lie within the bounds of variation for that ecosystem (White and Jentsch 2004). For example, grasslands that are heavily grazed by domestic or nomadic herds of *herbivores* will recover if they are grazed intermittently. If all grazing is halted, the system may well switch to a new trajectory. In ecological restoration projects, it is essential to distinguish between normal stressors needed to maintain ecological integrity and more intensive and damaging disturbances. Not all stress events protect ecosystem integrity, although spontaneous recovery from them is relatively rapid.

Moderate disturbance is more severe than that of a normally recurring stress event but insufficient to cause impairment. The result is a degree of damage that can be undone by an ecosystem's inherent properties of *resilience*. The process of resilience is a network of responses to recover ecosystem health and integrity. Human intervention is not needed, although it may accelerate recovery. As pointed out long ago by the late ecologist Walt Westman (1978), resilience can be measured in terms of the degree, manner, and pace of recovery of both structure and functionality of a damaged ecosystem as it returns to its former state after disturbance. Holling (1973) and Gunderson (2000) defined resilience first and foremost as the rate of self-recovery of an ecosystem from disturbance and, by inference, its capacity for self-recovery following disturbance. For our purposes, and in simple terms, if an ecosystem has sufficient resilience to recover from disturbance unassisted, and in what seems to be a "reasonable" period of time, there is generally no need to restore it. Although it is a bit complicated, let us note that resilience can also be said to describe the magnitude of disturbance that an ecosystem can absorb before it switches to an alternative state controlled by a different set of processes (Holling and Gunderson 2002; cf. Suding and Hobbs 2009, 5). For example, how much change in climate to hotter, drier conditions, can a forested ecosystem tolerate before it reaches a threshold of no return, and shifts to an alternative state, such as thicket or shrubland? Similarly, the impact of an invasive, alien species—a topic we return to later in this chapter and elsewhere—may appear to reach a point of no return, also known nowadays as a "tipping point." Restorationists will often encounter such situations in the years to come.

This alternative definition of resilience becomes relevant in deciding whether to select a restoration target that emulates a former, historical state at a site, or rather one that may be more likely to sustain itself under a new set of environmental conditions. If a restoration is directed toward a future state, it should be one into which an undisturbed ecosystem would likely develop in response to ongoing environmental change. In other words, the future state is anticipated from a continuation of the historic trajectory for that ecosystem in response to anticipated trends in exogenous conditions. In chapter 7, we discuss this topic in more detail.

Next, the third level of disturbance, impairment, is caused by severe damage, be it sudden or prolonged. Resilience is overcome in these cases, and ecological processes cease or are significantly altered, sometimes permanently. Biodiversity is altered. Natural (spontaneous) recovery, if it happens at all, is prolonged and is usually marked by significant demographic transformations, which may be expressed in stages consisting of seral (successional) communities. Ecological restoration is a specific antidote for impairment and may be the only option to reinitiate or accelerate ecological processes that induce recovery. There is no guarantee that spontaneous recovery or restoration will recover the preimpairment state.

A critically important area for exploration of these issues is the world's largest ecosystem, the deep sea. Compelling evidence is emerging (Ramirez-Llodra et al.

2011; Levin and Sibuet 2012) that the deep seas' biota and their habitats may have reached a tipping point caused by a combination of indiscriminant exploitation of fisheries (including unintentional side catches); scouring of benthic habitat by fishing vessels; discharge of sewage, other contaminants, and solid waste; and discharges from mining operations (Roberts 2002; cf. Roberts 2009). The problem may worsen as exploitation of petroleum and minerals increases, particularly at the deep end of continental margins, and greater depths as well (Levin and Dayton 2009). Impaired marine environments promise to open new opportunities and tremendous challenges for ecological restorationists.

## Degradation, Damage, Destruction

Now let us compare *degradation*, *damage*, and *destruction*, which are three somewhat overlapping ways that disturbance occurs. Degradation occurs when disturbance is prolonged rather than severe. It can result from chronic, low-grade disturbance, such as occurs when a natural range is overgrazed by domestic livestock over time. Degradation can also result from a punctuated series of minor impacts that occur at a frequency that disallows full recovery between impact events. For example, a forest can be degraded by harvests of individual trees at a frequency that exceeds their rate of replacement. The most desirable species are removed by this practice, which foresters call "high-grading" and what Arturo Burkart (1976) aptly called "artificial negative selection." The most desirable timber trees are progressively harvested and removed, thus leading to replacement by less desirable species of inferior ecological quality and timber value. Sadly, this has been a widespread practice by people around the world throughout history whereby the best portions of a resource are exhausted without thought for the future. The first human hunters in Madagascar killed off the biggest lemurs first. Newly arrived Aborigines, starting some 45,000 years ago, quickly killed the largest species of kangaroos for food in Australia. Chronic and punctuated disturbances weaken the ecosystem incrementally until the ecosystem reaches an ecological threshold beyond which it is unable—in a time scale of relevance to humans—to recover to its former state without assistance (Aronson et al. 1993a, 1993b). Intervention in the form of modified management practices, regulation to curtail overexploitation, and, very often, ecological restoration will be required.

In comparison to the preimpairment state of an ecosystem, alternative states of degradation are typically simplified in ecological complexity, depleted in terms of biodiversity, and incapacitated in respect to providing ecosystem services. Nonetheless, an altered state may be resilient on account of its dominance by a single, *invasive species*, such as *Cytisus scoparius* (Scotch broom) and *Imperata cylindrica* (cogon grass), which have the capacity to persist indefinitely. Therefore, it is important to avoid equating resilience as being a priori something good from society's perspective! Resilience is desirable, from a human point of view, if and only if the ecosystems at issue are yielding steady flows of ecosystem services to society, and appear to be capable of adapting to changing environmental conditions in ways that indicate that an evolutionary "engine" is at work.

Next, damage is the term applied to significant disturbance, or impairment, which occurs in a single, acute impact or disturbance event. Damage sometimes is sufficiently severe to cause an ecosystem to be transformed and switch to an altered state, assuming no restoration is undertaken. Investment in repairing the abiotic infrastructure of the damaged ecosystem during restoration is likely to be necessary as an intervention, prior to the introduction of biotic stocks. Damage, as well as degradation, is not exclusively a property of impaired ecosystems. Stress and moderate disturbance can also manifest as degradation and damage.

Destruction is the ultimate expression of disturbance recognized in the *SER Primer* (SER 2002). In the course of ecosystem destruction of a terrestrial system, essentially all organic matter is removed, including the living biomass of organisms and detritus: think of a backfilled, open-pit mine site. If recovery ever occurs, it will require recruitment or intentional introductions of seeds or other *propagules* for all plant species along with the recovery or replacement of soil and its biota. In the case of a destroyed aquatic system, redesign and engineering of the abiotic environment is generally paramount, after which most or all aquatic species will reappear by means of their inherent mobility, so long as connectivity to other aquatic systems has not been severed. Gully formation is a common form of destruction in a localized area (fig. 3.1); however, its detrimental effects can cover a much wider area on account of disruption to hydrology and sedimentation lower in its catchment.

The relationship between degradation, damage, and destruction can be illustrated as follows. A forest can be degraded by harvests of individual trees at a frequency that exceeds their rate of replacement. Alternatively, that same forest could have been damaged during a single event during which all trees were harvested (clear-cut). Damage is exacerbated if harvesting causes the soil to become destabilized and subject to erosion. This forest could have been entirely destroyed, if forest removal caused mass wasting (slippage) of a mountainside or if the soil was intentionally excavated to expose a seam of coal for extraction by opencast strip mining.

## Ecological Consequences of Impairment

What happens in an ecosystem that has been degraded, damaged, or destroyed? Or to one that has been intentionally transformed by people in the past, and now occurs in a state or condition that is considered undesirable? The answer is site dependent; however, some consequences can be predicted, particularly in terrestrial

**FIGURE 3.1.** Gully formation following the removal of a native ecosystem. To prevent future gullying, below-grade check dams of stones or other hard, natural materials may have to be installed at intervals before the gully is filled with soil and vegetated.

ecosystems. These include losses of specialized species and gains of generalist species; colonization by invasive species; simplification of community structure; microclimate disruption; losses of beneficial soil properties; reduction in the capacity for mineral nutrient retention; and alteration in the moisture regime. In short, impairment simplifies ecosystems. Here, we shall describe these consequences and three more—trophic cascades, *desertification*, and *salinization*—that may impact entire landscapes or regions, when human management of resources goes awry. Indeed, an important frontier for research and practice in restoration concerns the interactions among ecosystems: if one system is severely stressed, what are the impacts on adjacent or interlinked systems in the same landscape? Conversely, if one ecosystem is restored, can we expect cascades of positive effects in neighboring ecosystems?

### Losses of Specialized Species and Gains of Generalist Species

A radical change in vegetative cover is frequently an obvious indication of impairment and simplification. When an ecosystem is impaired, its vegetative cover tends to be replaced, at least in part, by fast-growing weeds, brush, or sometimes trees that were inconspicuous or absent prior to impairment. They do not require specialized habitat, and they tolerate unstable and physically disturbed site conditions. For those reasons they are called generalist species. These species tend to produce seeds or spores copiously and regularly. The woody species reproduce at a relatively young age and commonly produce small seeds and fruits. Generalist species forsake their competitive potential for an overwhelming reproductive capacity. MacArthur and Wilson (1967) categorized such plants by the term *r-strategists*, and such plants are often called *ruderals*, a word that originally meant plants that grow on rubble. An onslaught of generalist species following disturbance commonly increases species composition, which at first glance would seem to increase complexity rather than simplify an ecosystem. What becomes simplified, though, is ecosystem structure, which, along with losses of specialized plant species due to disturbance, greatly reduces previous ecological complexity.

In contrast, species of well-established, intact ecosystems tend to be competitive and persistent. They prefer stable conditions and may require specialized, well differentiated habitats. They tend to reproduce irregularly or at least not copiously. Instead, they dedicate their photosynthate (carbohydrate and other energy-rich compounds ultimately derived from photosynthesis) into vegetative structures that fortify their ability to withstand stress and competition and to increase their longevity. MacArthur and Wilson (1967) called the species typical of intact ecosystems by the term *K-strategists*, meaning that their life history is predominantly determined by the population carrying capacity (K) of a given environment.

Grime (1974, 1977, 1979) offered an alternative and complementary scheme of analysis in which he recognized three basic strategies by which plant species adapt to an environment, namely disturbance tolerators, stress tolerators, and competitors. He noted that most species display combinations of the three strategies. Grime defined competition as the tendency of neighboring plants to use the same volume of space and the same resources (water, nutrients, solar radiation). He defined stress as the various external constraints that limit dry matter production of plants, such as extreme temperatures; chronic deficiencies of water and nutrients; osmotic tensions from salinity, anoxia, and shade. Disturbance was the term Grime used for processes that reduce or destroy plant biomass by such means as mechanical damage from storms and floods, burning, herbivory, pathogens, mowing, and trampling. In light of Grime's C-S-R trilogy—usually drawn as a triangle and known as "Grime's triangle," for example, by Silverton et al. (1992)—we see that plants that adopt the disturbance tolerator strategy tend to occupy highly disturbed sites and qualify as r-strategists. Plants that Grime would have called competitors, as well as stress tolerators, tend to be K-strategists. The C-S-R classification can be useful in deciding the relative abundance of species that are to be planted at a restoration project site.

There is no reference document that lists K- and r-strategists or distributions of species among the C-S-R categories. These are plant strategies that become apparent with increasing familiarity with the natural history of a locality. Practitioners need to keep a watchful eye for evidences of these strategies among species at project sites. In ecological restoration projects, growers of nursery stocks tend to harvest seeds of r-strategists, because they can be collected in abundance and grown easily. Restorationists tend to plant r-strategists, because nursery stocks for them are readily available and inexpensive. Regulatory personnel tend to authorize the planting of r-strategists, because they are assured that regulated parties can obtain stocks of these species. Unfortunately, K-strategists are sometimes underrepresented at project sites to the detriment of ecosystem recovery. We advise that those who prepare restoration plans to establish of K-strategists insofar as possible and not expect these species to return spontaneously.

## Colonization by Invasive Species

Native species (also called *indigenous* species) are those that naturally occur in a location and are not known to have been transported there intentionally or inadvertently by human agency. Species (or infraspecific *taxa*) that were introduced to a location from elsewhere by human agency are called alien, nonnative, nonindigenous, introduced, or exotic species. These terms apply to plants and animals and also microorganisms. These newcomers commonly gain a foothold in their new haunts where competition from local species was compromised by disturbance.

An alien species is called invasive if it proliferates and noticeably replaces native species, usually in the absence of normal demographic controls that prevent its proliferation in regions where that species is native. Although proliferations by alien species have long been noted, the technical terms *invasive species* and the related term *biological invasions* are recent, and consensus is lacking for determining when a species warrants being designated as invasive. The recognition of invasive species is complicated by the propensity of native species to become invasive colonists themselves in impaired ecosystem, especially when they grow off site, that is, when they occupy habitat, or ecological niches, and positions in the landscape where they did not occur prior to disturbance.

All too often, the terms *alien* and *native* are used too loosely, that is, without adherence to biogeographical criteria. Plant and animal species introduced into the Western Hemisphere by European colonists may arguably be considered alien. However, the relevant date of 1492 has no significance in Africa or Eurasia. Within the Mediterranean Basin, endowed with more than ten millennia of biocultural melding, the determination of which species are truly native is purely conjectural in many cases and subjective in others (Blondel et al. 2010; Filippi and Aronson 2011a, 2011b). Some invasion science specialists, for example, Pyšek (1995), reach back 10,500 years ago and argue that a species is native to locations in Europe only if it occurred there prior to the *Holocene* or Neolithic period. However, if a species arrived subsequently, but without human activity, should it be considered native or not?

Many conservation biologists and environmentalists advocate the extirpation or control of all alien species, and particularly those known to be potentially invasive. With respect to ecological restoration, we are also concerned about biological invasions, because they arrest recovery in its early stages by co-opting space and utilizing ecosystem resources that would otherwise be available to those local species that we want to reestablish. In addition, alien and invasive species compromise ecological functionality and ecosystem wholeness.

A common demographic pattern is for an alien species to increase in abundance slowly without notice over a period of years or decades (called a lag-time), until it reaches a threshold population size. Thereafter, it increases rapidly in abundance (Richardson et al. 2000), and its populations seemingly "explode" (D'Antonio and Chambers 2006). For example, the Asian cogon grass was introduced into Florida and remained uncommon for many years. It was considered a botanical curiosity of no importance in the early 1980s. Before that same decade came to a close, cogon grass became strongly invasive and formed monospecific communities covering vast areas of disturbed land throughout much of the state.

Invasive plants commonly require control with herbicides, mowing, biological control or—for invasive animals—pesticides, shooting, and trapping. Otherwise they threaten the integrity of native ecosystems and alter ecosystem dynamics in profound ways. However, all of these measures—perhaps especially biological control—bring environmental dangers of their own.

Alien ruderal plant species may pose little threat to successful restoration, as long as they are not competing with native species and if they are insignificant in size, sparse, or appear only in marginal areas such as roadsides. However, in ecosystems where native r-strategists are characteristic and need periodically exposed open niches to persist, the occurrence of alien species becomes threatening. For example, VFT 2 describes Mediterranean steppe vegetation in France in which native annuals predominate. Of great concern are alien K-strategists that may block regeneration niches in this ecosystem and replace individuals or entire populations of indigenous species indefinitely. Intimate familiarity with the local vegetation by restoration practitioners is essential for identifying threats and knowing which alien species can be ignored.

Alien species issues near urban areas may preclude the possibility of performing ecological restoration. Birds feed copiously on fruits of alien plant species that adorn landscaped gardens in residential areas and disperse their seeds into adjacent natural areas. Native ecosystems can be overwhelmed by alien species that sprout from the continual rain of these seeds. For example, in Tallahassee, Florida, a forest dominated in every stratum by alien species (of *Cinnamomum*, *Ligustrum*, *Ardisia*, *Nandina*, etc.) replaced a native pine-hardwood forest within forty-three years (Clewell and Tobe 2011) (fig. 3.2).

Native species that sometimes behave as invasives are often vines and climbing shrubs that colonize in profusion following disturbance, and delay or block natural regeneration in disturbed areas. Examples from the southeastern United States include species of *Smilax*, *Vitis*, and *Rubus*; however, these are more realistically called nuisance species (Clewell and Lea 1990), because their competitive grip eventually dissipates without lasting effect. They retard, rather than prevent, recovery.

Other species that are unequivocally native can become invasive (Valéry et al. 2008, 2009; Simberloff et al. 2012). In English, we have no words to distinguish between these two kinds of invasives. To give an example, well-drained uplands with loam soils throughout the southeastern United States originally supported oak-hickory and pine-oak-hickory woodlands characterized by species of oaks

**FIGURE 3.2.** Forest dominated by camphor tree (*Cinnamomum camphora*) and other invasive exotic species in Tallahassee, Florida (Clewell and Tobe 2011).

(e.g., *Quercus stellata*, *Q. falcata*, *Q. velutina*, *Q. marilandica*), hickories *(Carya tomentosa)*, and shortleaf pine *(Pinus echinata)*, which were subject to occasional lightly-burning surface fires. Most of these woodlands were removed in the seventeenth, eighteenth, and early nineteenth centuries for intensive crop production. Thereafter, many agricultural lands were abandoned and were populated by seeds from the still-plentiful trees in moist lowland sites that were too poorly drained for growing crops. Forests of these lowland trees (e.g., *Pinus taeda*, *Liquidambar styraciflua*, *Quercus nigra*, *Q. laurifolia*, *Prunus serotina*, *Acer rubrum*, *Fraxinus* spp., and *Magnolia grandiflora*) became established offsite on abandoned agricultural uplands. These trees reproduce and maintain this lowland forest on upland sites. They exclude the natural regeneration of the original upland species on account of competition afforded by dense forest structure that discourages the spread of fire and paucity of seeds of upland species (Van Lear 2004). Creeping surface fire was originally an important agent of demographic control that kept trees of lowland species from colonizing uplands. Trees of lowland species that grow as persisting forests in uplands qualify as native invasive species, according to Brewer (2001) and Clewell (2011), and these forests should not be restored on upland sites. Interestingly, the pine-hardwood forest described by Clewell and Tobe (2011) was initially composed of these same native invasive species, and this forest was replaced within four decades by alien invasive species.

The area of the Earth's surface presently occupied by invasive species is enormous and has been increasing substantially each year. Invasive species are causing serious ecological disruptions globally to ecosystem structure and functionality. The problem mounts as introductions of new invasive species accelerate, and disturbed habitats become increasingly available for them. Invasive plant species can effectively block succession in large patches, owing to their competitive advantages. Some invasive animal species—of all sizes and shapes—can and do wreak havoc on native fauna. Examples include pigs, goats, rats, feral cats, the Indian mongoose, snails, bees, and snakes, as well as myriad microorganisms (Richardson et al. 2011).

Some ecologists and many less-specialized observers advocate that invasive species cannot anymore be controlled and that instead they should be accepted as the new biological norm. This topic is currently generating heated discussions (e.g., Davis et al. 2011; Simberloff 2011). Some dedicated and well-meaning conservationists, for example, Schlaepfer et al. (2011), suggest that nonnative species provide conservation values. Concern has been expressed by others (e.g., Simberloff et al., in review; Aronson et al., in review) that this does not mean we should abandon conservation and restoration practice, but instead take courage from the numerous successful campaigns undertaken to control or eradicate harmful invasives, and carry on in the same spirit, with all the means at our disposal.

Restoration practitioners are commonly faced with the decision of whether to allocate some portion of limited project funds to remove new and unanticipated invasive species or to ignore them and dedicate funds to the performance of previously scheduled restoration tasks. We advise restorationists to combat invasive species that clearly pose a threat and not to deplete budgets by trying to eradicate those that pose no immediate threat or for which no effective long-term control measures are available. Newly arrived species that are known to be invasive elsewhere should be watched closely and perhaps eliminated before they can become invasive locally. At a project site, however, restorationists may have to practice triage to decide which invasives can be effectively extirpated and which are too well established to even bother.

Recent investigations suggest that an invasive-resistant ecosystem could be restored by filling available niches with native species having similar ecophysiological and functional traits relative to those of potential alien invasive species (Funk et al. 2008). The idea is that native species would outcompete the invasive aliens. Testing of this theory would create an excellent opportunity for restoration ecologists to collaborate directly with restoration planners and practitioners. Where proven valid, it would be implemented in close collaboration between scientists and practitioners. An alternative strategy would be to manipulate the environment of the site undergoing restoration so as to favor native species and hinder the establishment of nonnative ones (e.g., Funk and McDaniel 2010).

Prevention should be the first line of defense against potentially harmful introductions of taxa; it is far more cost-effective than managing problems after an establishment, and it avoids occasional undesired side effects of management. The guiding principles on invasive introduced species adopted by the Convention on Biological Diversity in 2002 advocated a hierarchical approach: prevention as the priority response, early detection and rapid response and possible eradication when prevention fails, and long-term management only as the last option.

## Simplification of Community Structure

Structural damage to an ecosystem causes an overall loss of ecological complexity and ultimately the impairment of ecological function. As mentioned in chapter 1, community structure is usually a function of the sizes, abundance, spatial configurations, and life-form distribution of its species. In terrestrial environments, structure is largely determined by the prevailing plant species. In aquatic environments structure can sometimes be determined by sedentary animals, principally corals, oysters, and benthic in-fauna. In planktonic and nektonic communities, structure is not fixed or readily tangible but organisms that comprise these communities are still characterized in terms of their sizes, life forms, and abundance, and they tend to be juxtaposed in characteristic spatial configurations. Impairment simplifies and distorts structure. Species are lost or suffer reductions in abundance, particularly those that require specialized habitats. K-strategists tend to be replaced by r-strategists. Plants of a particular size class may be depleted, as happens when a storm uproots larger trees. Vertical stratification is usually reduced or simplified.

Ecosystem impairment may cause simplification in the spectrum of life forms and the functional groups they comprise. For example, we know a valley in northeastern India, in the Himalayas, where human intervention brought an oak forest to a tipping point through artificial replacement by a planted pine forest. Leaf litter consisting of pine needles accumulated to such depth and density that rainfall could no longer penetrate and percolate into the mineral soil. Instead, water cascaded over and within the litter to the stream below, leaving the mineral soil of the mountain desiccated much of the year. The pine litter posed a severe fire hazard—something that had not previously existed in this region.

## Microclimate Disruption

A microclimate is the meteorological deviation from of the prevailing climate (the macroclimate) caused by, and taking place within, an ecosystem. One obvious demonstration of microclimate is to walk into the refreshing coolness of a shaded forest on a hot day. The overall influence of an ecosystem is to modify or ameliorate extreme weather conditions of the macroclimate that prevails regionally. For example, wind velocity is reduced by vegetation and with it the desiccating effects of wind. Humidity increases accordingly, and moisture is retained in the ecosystem. In cool weather, heat is retained by that increased moisture. Incidences of frost and freezing temperatures are reduced. In hot weather, extreme heat is tempered by shade, transpiration, and evaporative cooling. Small differences in microclimate can have big effects on biota. The microclimatic reduction in wind velocity immediately above the ground in grasslands can be lifesaving for small organisms. Simplification of community structure reduces microclimatic effects and exposes organisms to the extremes of weather. Accelerated desiccation and exposure to frost are two common consequences, and both retard ecological function.

## Losses in Beneficial Soil Properties

Soils are complex and can be adversely affected by ecosystem impairment in a number of ways, including compaction, erosion, introductions of nutrients and contaminants, variations in electrical conductivity, and moisture availability. Soil compaction can be induced by heavy vehicles used in extractive processes or by repeated trampling by the hooves of domestic livestock. Machinery and livestock compress macropores created by tunneling shrews, annelids, and insect larvae. Precipitation is prevented from percolation into compacted soil and subsequent capture in macropores. Excess water runs off, causing greater amplitudes in the volume of stream discharge and a consequent depression in the water table between rainfall events. Compaction retards root penetration. Soil aeration is reduced, which, in turn, reduces the metabolism of aerobic soil organisms, including plant roots, numerous kinds of soil animals, fungi, and most bacteria. Other impacts on the soil can set off similar chain reactions that disrupt ecological function. Such impacts include erosion, sedimentation, leaching of nutrients, oxidization of organic matter, accumulations of salt, and the introductions of excessive nutrients from agricultural operations and pollutants from industrial discharge and urban runoff.

## Reduction in Capacity for Mineral Nutrient Retention

A common attribute of complex terrestrial ecosystems is their ability to sequester (capture) and recycle mineral nutrients. Soluble nutrients in the soil can be lost to the ecosystem by percolation of soil water to a depth below the root zone or by the lateral runoff of rainfall into streams. Such losses have been called leakage by ecologists. Leakage has at least two negative consequences for otherwise stable ecosystems. First, nutrient loss reduces the productivity of plants that are K-strategists. Such plants tend to sequester and accumulate mineral nutrients in their biomass and in the slowly decomposing litter that they produce. Some of these nutrients are released and exposed to leakage for brief periods before they can be reabsorbed by roots. If soils are damaged, exposure to leakage increases, soluble nutrients are lost, and the competitive ability of K-strategists to sequester nutrients is forfeited. Second, leakage makes nutrients available to opportunistic r-strategist plants that typically need ample and readily available sources of mineral nutrients. Damage to soils can impact soil fungi, many of which form mycorrhizal associations with the roots of vascular plants and translocate nutrients to them in exchange for carbohydrates. If these fungi are harmed and their extensive networks of nutrient-absorbing hyphae are reduced, their vascular plant *symbionts* will suffer from a corresponding reduction in nutrient supply. Losses of nutrients are an obvious problem. The residual effects from excessive agricultural amendments of nutrients, especially nitrogen, can also cause severe stress to ecosystems that are adapted to nutrient poor soils.

## Alteration in the Moisture Regime

Through self-organization, ecosystems commonly exhibit tight control of moisture recycling and the inputs and outputs of water (Descheemaeker et al. 2006). Most stable terrestrial ecosystems develop mechanisms to sequester moisture.

This can be accomplished by maintaining a favorable microclimate that moderates temperatures and reduces evaporation. It can also come about by the production of plant structures that capture precipitation and dew, by growth forms that retard percolation and runoff, by accumulations of humus and other detritus that absorbs moisture, and by the activities of soil organisms that loosen soil particles and create passages for retaining moisture. In other ecosystems, the opposite problem occurs: excess water accumulates that must percolate or run off to allow soil aeration. For example, *Juncus roemerianus* produces extensive mats of dense, superficial roots and rhizomes that resist penetration of water from frequent tidal inundation. The water runs off as surface flow, and the mineral soil and plant roots beneath remain aerated (Kurz and Wagner 1957). External impacts to ecosystems can reduce the effectiveness of moisture regulation and thereby reduce their productivity and stability.

Those who design and manage restoration projects should be aware of changes in water budgets as vegetation matures and succession proceeds at restoration project sites. For example, trees are efficient conductors of soil water and will lower water tables through transpiration by a meter or more in growing seasons (Trousdell and Hoover 1955; Sikka et al. 2003). The moisture content of soils may change substantially as trees planted at a project site grow and transpire increasing amounts of water in successive years. In addition, tree crowns intercept much precipitation that evaporates before reaching the ground. These variables in water budgets require consideration by planners of restoration projects.

## Trophic Cascades

Trophic cascade is a term referring to the destabilization of the normal demographic relationships among predators, prey, and vegetation, caused by reductions in the number of predators. Following release from predation, populations of herbivores expand and overconsume vegetation to the point that characteristic plant species are severely reduced and community structure is compromised. This is a widespread phenomenon in both terrestrial and aquatic environments, which can be addressed in restoration programs by reintroducing predators, restoring fragmented ecosystems that facilitate the movement of wide-ranging predators, and by other actions that promote the recovery of the missing predators. Eisenberg (2010) described her investigations of trophic cascading in North America, whereby the removal of wolves released elk from predation. The elk, in turn, prospered, and they overgrazed aspen saplings to the point that none could grow into trees. The lack of young aspen forests diminished populations of beavers and songbirds, both of which depended on young aspens for food and habitat. Eisenberg (2010) also reviewed the work of Paine (1966, 1969) on the removal of predaceous echinoderms (brittle star) from tidal pools, which allowed substantial increases in populations of mussels upon which it preys. Mussels, in turn, grew densely and preempted habitat needed by an array of marine plants and animals.

## Desertification

Many landscapes, especially but not only in less affluent parts of the world, have been mismanaged or exploited for agriculture or resource harvesting beyond their capacity for renewal, or they were ravaged ecologically by warfare and armies of invading tribes and empires. A common consequence is loss of organic matter from soil, which reduces cation exchange capacity, facilitates nutrient leaching, and hampers retention of soil moisture. Soil compaction or erosion follows and increases rates of surface runoff from precipitation. In response, such landscapes become drier and less productive, a process called *desertification*. The process creates land that is more arid or xeric than before but not necessarily a desert. This trend is generally irreversible unless an intensive, expensive, long-term restoration effort is undertaken.

Desertified regions may receive less rainfall and dew because of a reduction in water vapor that was previously generated from evapotranspiration. Aquatic ecosystems suffer from desertification occurring elsewhere in a catchment. Precipitation is no longer retained on the land and instead moves rapidly into streams via surface runoff. The amplitude of stream discharge increases, with surges or spikes of discharge in wet seasons followed by prolonged periods of low water flow. Discharge is more likely to be turbid from suspended particulates that erode off the land. The discharge will carry more nutrients from the land, which causes eutrophication in receiving waters. Pulsed stream flows into estuaries can cause stressful changes in salinity to marine organisms.

Desertified environments pose serious challenges to restoration practitioners. The restoration of desertified lands has been attempted (with small success, unfortunately) in Africa and elsewhere. New and improved approaches to the restoration of desertified lands are becoming available and more widely applied (Bainbridge 2007; Tongway and Ludwig 2011). Full recovery of badly degraded ecosystems in arid and semiarid regions, though, may never fully occur under existing climates. Restoration of desertified landscapes is increasingly becoming incorporated as provisions in international treaties. The UN Convention on Combating Desertification calls for ecological restoration, as does the Convention for Biological Diversity.

## Salinization

Another form of desertification is salinization, which has occurred on a vast scale in dry lands throughout the world (Dregne 1992; Kassas 1995; Rengasamy 2006).

Salinization commonly occurs when croplands are flooded by irrigation water, some of which evaporates and leaves a residue of dissolved salts on the soil surface. Repeated irrigation causes increased electrical conductivity and consequent toxicity to plants.

Salinization also develops in those arid areas where salts are dissolved in groundwater that lies below the rooting zone of plants. If there were sufficient rainfall, the soil would become moistened to a depth that reached the saline groundwater, and the salt would diffuse upward to the soil surface. Ordinarily, low rainfall, in combination with high rates of transpiration of soil moisture by deep-rooted trees, prevents rainwater from penetrating deep enough in the soil to reach saline groundwater. If trees are removed, rainfall may continue to percolate downward until it reaches the saline groundwater. The dissolved salts are then able to diffuse gradually to the surface, where the water evaporates and leaves a saline residue. The process continues until the salt content at the surface becomes toxic to plants, causing what is called a salt scald. Deposits of these salts are highly mobile in surface runoff and can spread their toxicity to lower-elevation lands. This salt may reach streams, where it raises salinity and affects aquatic ecosystems.

Many people have fled landscapes that became desertified, salinized, or otherwise degraded from exploitation and have taken up lives as what can only be called "ecological refugees," in urban slums of cities such as Lagos, Nigeria, Mumbai, India, and Mexico City. Those remaining in rural areas intensify desertification as they fight to scratch out a living. To prevent such a downward spiral, a public-private coalition formed in southern California to prevent water used for drinking and irrigation from becoming increasingly saline (see http://www.socalsalinity. org/). Ecological restoration, in concert with a program for the *restoration of natural capital* (RNC) that includes such a public-private coalition, may represent the most effective strategy to prevent and perhaps reverse large-scale degradation (Aronson et al. 2007a; chaps. 9 and 12).

David Printiss

Seemingly endless expanses of pine savannas once covered most of the coastal plain of the southeastern United States from the Carolinas south to Florida and west to Texas. These floristically diverse, fire-maintained ecosystems occurred across an ecological landscape that ranged from xeric sandy uplands (called sandhills or sand ridges) to bogs that become shallowly inundated in wet seasons. Only 2 or 3 percent of pine savannas still exist, and many stands suffer from altered fire regimes, which allowed colonization by woody species typical of plant communities that are not fire maintained (photo 1).

**PHOTO 1.** Xeric sandhill and second-growth longleaf pine savanna with wiregrass-dominated undergrowth. This is a reference site and an undergrowth seed-donor site that lies between forested steephead ravines on either side.

Most of the plant communities comprising the southeastern pine savanna biome are (or were) characterized by open stands of widely spaced longleaf pines (*Pinus palustris*) growing within a dense multispecies and grass-dominated groundcover. Wiregrass (*Aristida stricta*)—a long-lived bunch grass that provides much of the fuel for frequent surface fires—prevails in pine savannas in Florida. Approximately thirty to eighty additional species per hectare grow interspersed with wiregrass, more on wetter sites and less on drier sites. These include grasses, sedges, other herbs, and low-growing shrubs. A few potential small trees are interspersed and typically grow as coppice not much taller than the herbaceous cover before they are pruned back to their root-crowns by frequent fires. Two of the most common trees found in xeric sandhills are turkey oaks (*Quercus laevis*) and sand-live oaks (*Q. geminata*).

Many former xeric pine savannas in northern Florida have been converted to industrial timber lands. One such region flanks the eastern side of the Apalachicola River in northern Liberty County, between Bristol and Torreya State Park, Florida. The mature, original growth of longleaf pine was clear-cut, probably in the 1930s or earlier. The wiregrass-dominated groundcover remained intact, and young pines regenerated. In the late 1950s, these second-growth pines were cut. The groundcover, including wiregrass and logging debris, was removed by root rakes and windrowed in preparation for industrial offsite slash pine planting for pulpwood production.

This region has long been known for its unusual geological and biogeographic features. Its sands are deeply incised by steep ravines formed by an unusual process called sapping erosion. Abundant precipitation percolates through sand until it is diverted laterally by a clay layer ~ 50 meters below the surface and emerges as a freshwater spring at the heads of ravines. Sand carried by springs continuously erodes ravine heads ("sapping") forming what are known as steepheads. Steephead ravines, which occur in only a few locations globally, provide habitat for an exceptionally diverse, mixed hardwood slope forest that contains rare and narrowly endemic species, including two conifers: Torreya tree (*Torreya taxifolia*) and Florida yew (*Taxus floridanus*). These and other rare species were once part of an Appalachian mountain flora that was pushed southward in the Pleistocene and persist as relict populations in moist, relatively cool ravines (photo 2).

The industrial slash pine plantations failed in the exceptionally dry and infertile soils. The Nature Conservancy (TNC) purchased 6,295 acres of this land, which is called the Apalachicola Bluffs and Ravines Preserve (ABRP), beginning in 1982 in order to preserve and protect the slope forests and seepage streams in the ravines. Not long after acquisition it became apparent to TNC personnel that the adjacent sandhills and the xeric longleaf pine savannas they supported were integrally related to the ecological health of the

**PHOTO 2.** View from sandhill into steephead ravine into which a tongue of fire had recently burned.

slope forests. For example, tongues of fire occasionally burn from sandhills into the ravines and maintain habitats that harbored some of the region's endemic species. What began as a relatively simple land protection project grew into a large-scale sandhill restoration project.

Prior to the onset of the restoration at ABRP, very little was known concerning the tools and methodology necessary for xeric pine savanna restoration, and we did not even know how to manage wiregrass so that it produced viable seeds. Accordingly, the first ten years of the project were dedicated to testing restoration techniques to discover which ones could be used effectively on large areas at acceptable cost. During the mid-1990s, a small, dedicated staff was gathered and equipped to apply what had been learned. Although much has changed in subsequent years regarding operations and equipment, the basic findings of initial restoration efforts were confirmed: frequent fires are critical, and a dense groundcover consisting mostly of grasses is needed as fuel for these fires.

Fortunately, a few tracts of land at ABRP were not root raked. The tracts were used as reference sites and as groundcover seed donor sites for restoration. Elsewhere on the property, windrows were leveled with a bulldozer, and the residual sands and topsoil were spread across the land in preparation for restoration. A bulldozer eliminated offsite hardwoods (e.g., *Quercus hemisphaerica*, *Q. nigra*) and also reduced the abundance of upland oak species (e.g., turkey oak and sand live oak) that had survived industrial forest operations and proliferated during the past several fire-free decades. Some hardwoods were girdled and treated with herbicide. This seemingly heavy-handed approach was necessary to reduce competition and favor desirable species reintroductions. Stunted offsite slash pines (*Pinus elliottii*), were thinned but not entirely removed so that their combustible leaf litter would contribute fuel once burning resumed.

Groundcover donor sites were burned in the growing season, especially in May, to stimulate seed production in wiregrass and associated groundcover species. Seeds were harvested in November and December of the same year, using FlailVac® and Prairie Harvester seed collectors mounted on all-terrain vehicles and tractors. As soon as practical (usually January or February) the seed mix was planted at the restoration site using a Grasslander® planter. Seeds fall from the hopper of the Grasslander into furrows it makes, which are closed immediately by pressure from rubber tires, ensuring tight contact of seeds with the mineral soil. The Grasslander was retrofitted with spring tine harrows rather than disks to form furrows, because tines were able to negotiate residual woody roots, stumps, and debris better than disks. Seeds of wiregrass sprouted during rainy periods over the course of the next twelve months following planting. Immediately following the sowing of the groundcover seed mix, containerized longleaf pines were planted at a rate of 615 trees per hectare. That density is high for this plant community and may eventually require thinning. However, these pines produce resinous leaf litter that adds considerably to the flammability of the surface fine fuels (photos 3 and 4).

Forty months after planting, the wiregrass-dominated groundcover was burned. At the time of burning, some planted longleaf pines had grown to a sufficient height to survive fire, and others remained in the grass stage (rosettes without aerial stems) and readily survived fire. A minority of the planted pines were fire killed. Their loss reduces the need for thinning pine populations later, and the loss of flammability contributed by their needles is replaced by surviving trees that produce an increasing biomass of needles as they continue to grow. At this point, most restoration tasks have been successfully completed, and subsequent activities will largely be considered under the aegis of ecosystem management (photo 5).

**Photo 3.** Three, tractor-drawn Grassland® planters planting undergrowth seed mix on a newly prepared project site.

**Photo 4.** Alternative Spring Break volunteers plant longleaf pine seedlings immediately after seeds of undergrowth plants are planted by the Grasslander® seen in the background.

**PHOTO 5.** Restored xeric longleaf pine-wiregrass savanna extends to the horizon. The wiregrass-dominated groundcover was burned 40 months after planting, and young longleaf pines survived fire (the larger trees are ~ 15 years old). The photo was taken several months later, after the wiregrass had recovered from fire and was producing seeds. Compare with photo 1.

The earliest trials of restoration focused on raising wiregrass plugs in a nursery and outplanting them at restoration sites. Methods using this approach were successful but eventually proved to be cost prohibitive (Seamon 1998). The first direct-seeding protocol was highly successful: initial sampling of direct seeding in 75, 225 square meter plots revealed a mean of 48.5 to 66.3 plant species in the study plots, indicating recovery of species composition. However, it depended on several independent pieces of equipment and was labor intensive (Cox et al. 2004). These early attempts were funded on minimal budgets and were fueled by the ingenuity of TNC employees and the dedication of numerous volunteers from Tallahassee and surrounding areas.

The current direct-seeding method using Grasslander planters was developed in the late 1990s and has proven to be both dependable and efficient. Reconnaissance and comparisons to the results suggest similar success in recovering undergrowth. It has been used by TNC personnel to restore 200 acres of longleaf pine sandhills per year at ABRP and on adjacent property for Torreya State Park for the last seven years at an acceptable cost of $1,250 to $1,700 per hectare, including site preparation, seed collection, direct seeding, and pine planting.

Much of the manual labor for the restoration project using the current direct-seeding method has been supplied by volunteer inmates from a local correctional institution. The labor crews consist of ~ 8 inmates and an inmate supervisor who was trained by TNC personnel in restoration methods. Their tasks include pine planting, some hardwood removal, and coarse cleaning of the seed mix. The inmates are treated with respect and allowed to use most of the same tools as other volunteers and staff. It is commonly known that inmates strongly prefer to work on a TNC crew (rather than on work details elsewhere), in spite of having to spend the day beneath the unrelenting Florida sun.

Until the current technique was developed, it was the general opinion that xeric longleaf pine-wiregrass savannas were beyond the reach of ecological restoration, if not technically then at least in terms of cost. If so, then large regions on the Atlantic Coastal Plain could never be restored. The present technique, however, has demonstrated that such restoration is both technically and economically feasible. Importantly, there should be little or no need or cost for subsequent ecosystem management, other than prescribed burning, which is needed on all southeastern pine savannas, including those that were never disturbed.

Thierry Dutoit, Renaud Jaunatre, and Elise Buisson

La Crau steppe in southeastern France is a xeric, semicultural landscape that was shaped by the interactions of sheep grazing with soil and climate since Neolithic times (photo 1). These interactions were responsible for temporal and spatial variations in the organization of the plant community from local to regional scales (Henry et al. 2010). The climate is Mediterranean with an average of 540 millimeters yearly precipitation, mainly in spring and autumn. The soil is approximately 40 centimeters deep, and half of its volume consists of siliceous stones. The soil overlays calcareous conglomerate that plant roots cannot easily penetrate. The plant community is species-rich (more than 40 species/square meter) and is composed of 127 species, mainly annuals. It is dominated by a perennial grass (*Brachypodium retusum*) and thyme (*Thymus vulgaris*). This community is unique in Europe. It serves as habitat for numerous steppe birds, such as the pin-tailed sandgrouse (*Pterocles alcata*) and the little bustard (*Tetrax tetrax*); and two endemic insects (a beetle *Acmaeoderella cyanipennis per-*

**PHOTO 1.** The reference, a semicultural steppe after several decades of grazing but no plowing. In the foreground are the characteristic stones of La Crau bearing mosses and century-old lichens.

*roti* and an endangered grasshopper *Prionotropis hystrix rhodanica*).

Large areas of La Crau steppe have been allocated for cultivation since the 1600s, reducing its original area by 80 percent (Buisson and Dutoit 2006). Future development threatens the remaining area. To avoid further losses, the Biodiversity Division of the governmental CDC Bank, the French Ministry of the Environment, and various regional government agencies joined to finance a pilot ecological restoration project in 2008. This pilot project will facilitate planning for future restoration projects that will be required to compensate anticipated environmental damage in advance of development. An industrial orchard, which had been established in 1987 near the center of the steppe and was abandoned in 2006, was chosen as the pilot project site. This 357 hectare site was purchased so that several restoration techniques could be compared experimentally over large areas. The reference model for restoration was the ecological description of the La Crau steppe published by Molinier and Tallon (1950). In 2008, vegetation and soil characteristics were inventoried to document the impaired site prior to restoration. In 2009, 200,000 fruit trees and 100,000 poplar trees, the latter having been planted as a windbreak for fruit trees, were cut down and removed. Thereafter, the soil was leveled.

The two-year objectives of restoration project work were to establish characteristic steppe species and limit the colonization of competitive ruderal weeds (species of *Chenopodium, Amaranthus, Bromus, Silybum,* etc.) (photo 2). In order to reestablish traditional cultural practices, local sheep breeders were engaged to reintroduce grazing flocks in spring 2010. An early benefit of grazing was control of ruderal weeds (photo 3). The objectives for the longer term (> 10 years) will be to redirect the plant community on the desired successional trajectory toward the reference steppe in terms of plant species composition and richness and community structure.

Intentional natural regeneration of the plant community was not an option for restoration, at least not in the short term, because of the low potential for seed dispersal by target species and competition from ruderal weeds that respond to residual increased fertility on the formerly cultivated soils (Römermann et al. 2005).

Four large-scale restoration treatments were attempted:
- nurse species establishment
- topsoil excavation and removal
- direct seeding of native species
- amendment of salvaged donor topsoil

**Photo 2.** Invasion by ruderal plant species (here, *Silybum marianum*) at the project site several months after trees were removed.

**Photo 3.** Same view as photo 2, one year after the reintroduction of sheep grazing. The plant community is now dominated by common grass species.

For the nurse species treatment, a seed mix of *Lolium perenne* L., *Festuca arundinacea* Schreb., and *Onobrychis sativa* Lam. was sown. These species were chosen for their palatability to livestock, their availability in local plant nurseries, and their ability to rapidly cover bare ground. The topsoil removal treatment consisted of excavating and removing the nutrient-rich upper soil layer to a depth of 20 centimeters. For direct seeding, seeds were gathered by air-vacuuming (photo 4) in the summer of 2009 and immediatly scattered by blowing them on the topsoil of the restoration site (1.5 kilograms per hectare) without any particular soil preparation. Soil for the soil amendment was salvaged from a donor site consisting of undisturbed steppe that had already been designated as a future construction area. The upper 20 centimeters of soil was gathered in bulk, transported, and spread the same day in the abandoned orchard in autumn 2009, a few hours before a significant rain. In order to preserve the genetic integrity of local populations, seeds and soil amendment materials were gathered in areas that had been dedicated to sheep grazing from Neolithic times until the establishment of the orchard in 1987. One area at the project site was not treated and serves as a control.

We evaluated the four treatments in 2011, two years after their application. Areas that were amended with salvaged soil from the donor site yielded favorable results. Species richness was very similar to that of the reference

**Photo 4.** Workers gathering seeds with vaccuums from a nearby intact steppe.

steppe. Most reference species (107 of the 127 known steppe species, or 84 percent) were recorded at least once over the treated areas with the exception of three of the most common species: *Brachypodium retusum*, *Thymus vulgaris*, and *Asphodelus ayardii* (Jaunatre et al. 2012).

Direct seeding resulted in the germination of a dozen desirable species from the reference model, whose plants were unevenly distributed (photo 5). These steppe species have the potential to increase in abundance and become more evenly distributed, as long as competitive grasses are controlled by sheep grazing.

Areas that were treated by sowing nurse species and by removing excavated topsoil yielded low values for species richness and species similarity relative to the reference model. However, desirable species richness was consistently higher in all four treatment areas than it was in the control plots. Nurse species inhibited colonization by undesirable ruderal species. Topsoil removal also removed the seedbank for ruderal species and, of course, any desirable species that may have still been present. The results of all four treatments will be evaluated for their potential applications in subsequent restoration projects in La Crau steppe.

**Photo 5.** Restoration project site 2 years after sowing of native seeds.

Bob Stewart and Tanya Balcar

The Western Ghats—a rugged mountain range that runs along the western coast of India—protrudes eastward in south-central India and becomes the Palni Hills. The original vegetation of this mountainous highland consisted of tropical evergreen forest called shola. The trees are massive in girth but stunted, presumably due to elevation at which they occur (1,500–2,400 meters), and they support diverse undergrowth. Sholas, which presumably were once widespread in previous geological eras, are now confined to valleys and ravines. Prior to recent decades of land use, more than 75 percent of the Palni Hills were covered by grasslands, which had become established following the glacial epoch and consisted of species with Himalayan affinities. The term *shola* is commonly applied to both the evergreen forests and their associated grasslands as an ecological landscape (photo 1).

Beginning circa 1830, the shola grasslands became increasingly exploited as farmland for growing potatoes and for grazing by domestic cattle and sheep. Exotic tree plantations were established and gradually replaced the

**PHOTO 1.** Intact shola landscape consisting of grasslands and patches of evergreen forest nestled in ravines.

grasslands. These trees reproduced and became invasive, including pines (*Pinus patula* from Mexico, *P. radiata* from California) and Australian *Eucalyptus* (especially *E. globosus*) and *Acacia* (especially black wattle, *A. mearnsii*). Black wattle is a small, short-lived tree that was introduced for its tannin. All of these exotic trees are phreatophytes, that is, trees with deep roots that exploit groundwater year around, in contrast to grasses of the sholas, which are dormant during the January–March dry season. As the grasslands were undergoing reallocation to tree plantations, shola forests were being decimated for lumber and domestic cooking fuel.

The small city of Kodaikanal (10°12'N77°28'E; 2,100 meters elev.) in the Palni Hills was a popular summer retreat of the British during the time of colonial rule and later became a year-round tourist destination. We settled there in 1986 and soon established a tree nursery. Vattakanal Conservation Trust (VCT), a local nongovernment organization (vattalkanalconservation-trust.org) was formed in 2001 to recover sholas and to contribute in a modest way to local betterment. Concurrently, young men from the city had organized through the auspices of an outside social agency, and they embraced the opportunity to help us in our work. We established the Vattakanal Tree Nursery to supply black wattle trees for domestic fuel, so as to reduce the need for cutting trees in shola forests. We also grew native shola trees to help regenerate damaged forests.

Another concern was the drying up of the watersheds on account of excessive transpiration caused by exotic tree plantations on former grasslands. Another organization, the Palni Hills Conservation Council, had produced a documentary film that dramatized the relationship between exotic tree plantations and diminishing water supplies. We showed that film, which stimulated many more residents of Kodaikanal to join in our efforts. Tree planting became a popular activity. Frequent newspaper articles encouraged these volunteer efforts. In 1993, we arranged for volunteers to spend several days in another community where similar conservation efforts were under way. The occasion served to raise awareness that local initiatives in Kodaikanal were part of a larger movement. This experience helped to stimulate the development of Vattakanal Organisation for Youth, Community and Environment (VOYCE), which soon attracted charitable donations from student groups in the United Kingdom. Young women became involved, some of whom became proficient in plant identification and the preparation of herbarium specimens. The fruits of our efforts multiplied to protect sholas. School children at a festival performed a play called "Don't Cut the Trees."

A local garden club formed with UK support. The club was a focal point of civic pride and became intertwined with forest stewardship.

In the meantime, the state forest department of Tamil Nadu, which administers public lands in the Palni Hills, became concerned with diminishing public water supplies and realized that transpiration from exotic plantation could be a contributing factor, although technical opinion was divided on that issue. The forest department decided to improve grazing lands for wildlife, including the 37-hectare Vattaparai catchment, 5 kilometers from Kodaikanal. The VCT entered into an agreement whereby the forest department would remove trees. The VCT would undertake the restoration of the grassland and monitor groundwater. This former grassland was densely covered by black wattle and some eucalyptus trees. Four narrow streams drained the upper slopes. They coalesced into a single channel that broadened into a small (~ 0.5 hectare) marsh as the gradient diminished at the base of the catchment. Since the site had become forested, the streams and marsh were usually dry, presumably due to excessive transpiration from exotic trees. Trees had partially colonized the marsh, and it contained insufficient water to support wildlife. Prior to the initiation of our project, the forest department had excavated a small pool within the marsh and lined it with concrete to capture water that wildlife could access (photo 2).

**PHOTO 2.** Pool excavated in dried-out marsh to provide water for wildlife.

In 2005, approximately 5 hectares of forest were removed on either side of the stream and in the dry marsh. VCT engaged Michelle Donnelly from Australia to establish groundwater-monitoring wells on the newly cleared land, one on either side of the marsh and four others at increasing elevations along the course of the stream.

The VCT operates on an annual budget of $3,250 to conduct monitoring, foster grassland and marsh reestablishment, and wage a constant fight with the emergence of black wattle from a well-stocked seed bank in the Vattaparai catchment. Children from two local schools, as well as shopkeepers from the nearby tourist spots, have removed many wattle seedlings. Seeds of native grassland and marsh species were collected, grown in our nursery, and outplanted in 2009, 2010, and 2011 (photos 3, 4). Grasses that were planted included *Arundinella vaginata*, *Botriochloa foulkesii*, *B. kuntzeana*, *Brachiaria semiundulata*, *Chrysopogon zeylanicus*, *Eragrostis schweinfurthii*, *Eulalia phaeothrix*, *E. wightii* (a Palni Hills endemic), *Heteropogon contortus*, *Ischaemum indicum*, *Setaria pumolla*, *Themeda sabrimalayana*, *T. temula*, and *Tripogon bromoides*. In 2011 we outplanted nursery-grown stocks of the

**PHOTO 3.** Shola grassland restoration-project site after forest removal. A small stream, tree stumps, and white casings for two water monitoring wells are visible. Photo courtesy of M. Donnelly.

**Photo 4.** Grass planting. Black wattle grows densely at the edge of the clearing.

spectacular and densely low-growing grassland shrub, *Strobilanthes kunthiana* ("kurinji," Acanthaceae). Some native grasses, forbs, and shrubs colonized spontaneously (e.g., the showy gentian, *Swertia corymbosa*). Others had persisted in competition with black wattle (e.g., the tree fern *Cyathea nilgirensis*). Marsh species that colonized included *Athyrium solenopsis, Bulbostylis densa, Carex speciosa, Coelachne simpliciuscula, Eleocharis congesta, Eriocaulon nilagirense, Fimbristylis* sp., *Hydrocotyle sibthorpioides, Hypericum wrightianum, Hypoxis aurea, Isachne angladei, Juncus effusus, J. leschenaultii, Lipocarpha chinensis, Mariscus cyperinus, Pycreus (Cyperus) flavidus, P. sanguinolentus, Rotala rotundifolia, Schoenoplectus mucronatus, Sciprus fluitans,* and *Utricularia graminifolia.* We grew *Carex longipes* from seeds and planted it in the marsh. By 2008, we had identified 85 plant species at the project site; they numbered 122 by 2011. We are fortunate to have available *The Flora of the Palni Hills* (K. M. Matthews, Rapinat Herbarium, Tiruchi, 1999), which serves as a valuable secondary source of reference information for this highly degraded region.

We have to maintain vigilance on domestic cattle, which can damage our new plantings and cause soil compaction. Some exotic grasses have yet to be controlled, especially *Pennisetum clandestinium* (East Africa). Removal of wattle from steep gradients has required repairs of incipient erosion problems.

Since tree removal in 2005, ponding in the marsh has increased each year and now contains water year around. In 2011 we observed groundwater upwelling from the floors of the marsh. Two of the streams leading into it flow all year. Monitoring data await technical analysis, but field observations have shown an immediate response in terms of increased groundwater availability following tree removal from the 5 hectare project site. Fish populate the marsh, and dragonflies are seasonally abundant.

We now have resident populations of the Indian wild bison (*Bos gaurus*). In a spectacular instance of predation and ecological function in 2010, Dholes (wild dogs) chased a young bison into the marsh and overcame it. Wild boars are in residence, and Sambar deer have visited the site. Birds that have appeared after the project began include peregrine falcon, blue kingfisher, paddyfield pipits, and scimitar babblers.

The restoration effort continues with hopes that the forest department will facilitate the removal of additional tree cover. So far, the partnership has been successful between the forest department and local community organizations, especially the coalition of VOYCE and VCT. Personnel from the forest department officially participated for a day in grassland restoration in 2008, and professional employees of the department sometimes contribute their personal support to the project on an individual basis.

# PART II

# What We Restore

Part 2 addresses what it is, exactly, that we are trying to restore. It is a lot more than meets the eye, and it requires us to "think like an ecosystem," which is not an everyday occupation. Recovery from impairment is the obvious intent of ecological restoration, yet ecosystems are in large part the products of disturbance, both human and nonhuman mediated, so we need to tease apart the notions of disturbance, stress, and impairment. To begin, chapter 4 attempts to make sense of this paradox and explains that *recovery* is a highly nuanced concept with multiple meanings. In chapter 5 we get to the heart of ecological restoration as we understand it—a truly holistic endeavor. There we identify the eleven attributes that collectively make an ecosystem whole. Chapter 6 returns to the theme of disturbance, but this time from the perspective of cultural activities that sometimes lead to ecosystem wholeness—another paradox. The key term there is *disturbance regime*—something rational and balanced, like a good diet—the opposite of a binge.

# Chapter 4

# Recovery

According to the SER (2004) definition, ecological restoration assists recovery. What do we mean by recovery? Recovery is a deceptively simple term for a surprisingly complex concept. If we recovered an antique automobile for display, it could remain in its refurbished condition indefinitely for all to admire as a gleaming legacy of our industrial past. Unlike a car that was constructed of inert materials, an ecosystem is different. It consists of living organisms that are constantly reacting to each other and to their abiotic environment. As stated in chapter 1, an ecosystem can be restored to its historic trajectory, but it can't be restored as a replica of a prior state. The reason is that its component plants and animals can't be fixed in time as if they were pasted onto herbarium sheets or preserved in formalin. Species populations undergo continuous changes in demography—germination or births; deaths; growth and reproduction; migration; evolution and extinction. Ecosystems are constantly changing, at least subtly, with every blink of an eye.

Restoration and recovery have different meanings, but they are frequently used interchangeably. According to Webster's Third New International Dictionary of the English Language, Unabridged (1993), restore means to put back or bring back to a former or original state, to rebuild or renovate. Recover means to take back or bring back to normal balance or self-possession—to rescue, cure, heal, or retrieve. According to these definitions, restoration means the return to a former state, whereas recovery means return of former potential and thus anticipation for subsequent activity. With these meanings in mind, you could restore an antique car but not recover it. Conversely, you could recover an impaired ecosystem and bring it back to normalcy or readiness to resume ecological processes, but you could not restore it! We presume it is far too late to suggest that we change the name of our discipline to Ecological Recovery. Moreover, the word *restoration* has proven itself to have enormous value and resonance in languages around the world. Therefore, let us go forward.

The SER definition of ecological restoration (SER 2004) states that we only "assist" recovery. This stipulation reminds us that the ecosystem plays a vital role in its restoration. Our intervention supplies the conditions for restoration, but human agency is only one factor in the restoration process. If we were restoring an automobile, we would be engaged in every aspect of restoration until it was ready for display. In ecological restoration, nearly all recovery is accomplished by living organisms. As restorationists we serve only as facilitators for those plants, animals, and microbial forms to perform the recovery. If we became more involved, it would not be ecological restoration. Instead, it would be gardening, landscape design, agronomy, *ecological engineering*, or some other activity whereby we would dictate the final ecological outcome or endpoint within relatively narrow limits. We would produce a version of an ecosystem that matched our conception of what we thought Nature should look like or what would fit our budget and time constraints. The ecosystem would necessarily be *anthropogenic*, in spite of the fact that it consisted of living organisms and superficially appeared to be "natural." We would have selected those species and arranged them according to our own design and our particular conception of Nature. Restoration uses tools from gardening, agriculture, and ecological engineering in assisting recovery. But what distinguishes restoration from these disciplines is our willingness to let an ecosystem evolve according to its own inherent properties. Once the patient has been assisted sufficiently to walk again, he/she chooses the direction in which to travel.

As facilitators in assisting recovery, we are limited in what we can do. We can prevent the causes of impairment. We can make corrections to the biophysical environment in order to facilitate the resumption of ecological processes. We can facilitate normal exchanges of organisms and materials with the surrounding landscape. By intervening in this manner, we are essentially returning an ecosystem to a position where it can recover by means of its own internal processes. Our actions may require great skill, arduous effort, and subtle acumen, but these efforts boil down to the repair of impacts to the physical environment, the reintroduction and reestablishment of missing species, and the removal or control of undesirable species. If we compared this level of effort to that of restoring the antique automobile, it would be equivalent to taking the car to a garage, striping its worn out parts, and placing new parts and various tools on the workbench. The car would reassemble itself and honk its horn to let us know when it was finished.

The product of ecological restoration is necessarily openended and indefinite. As humans, we want to control our environment. As restorationists, we relinquish our inclination to control natural processes and landscapes, so that a restored ecosystem will be as natural as possible. Thus, despite similarities, and some overlap, the approach of ecological restoration differs from that of engineers, farmers, and landscape designers who specify the end products of their work and ensure that they are attained. Ecological restoration is more akin to raising children. We can nurture them and control their behavior at first, but as time progresses we must release them and let them find their own way. We may have an idea of how they will turn out, but we may be surprised and we won't know for sure until it happens.

We limit our assistance, so that the impaired ecosystem recovers to a state of naturalness equivalent to that of the preimpairment ecosystem. This ideal can be approached but not fully achieved for two reasons, as explained in chapter 1. One reason is that the entire biosphere is already strongly affected by human influence, including anthropogenic introductions of alien organisms far from their centers of origin and global dispersal of industrial contaminants. Therefore, while restoration attempts to address these problems, some may be insurmountable for budgetary reasons, or because resolving them is beyond the scope of our current ecological ability. We may have to live with an invasive species, for example, still present but hopefully controlled, in a restored system. Second, many and perhaps most ecosystems are semicultural, and human activity must be counted among the environmental factors that shaped and maintain them. In addition, any intervention performed by a restoration practitioner necessarily introduces a "founder's effect" that may never entirely dissipate, even when the intervention consists of reintroducing extirpated species. When we say that we are restoring natural ecosystems or restoring an ecosystem to its natural condition, we do so with these caveats in mind.

After it is determined that a restoration project is feasible, and the sponsoring organization decides to move ahead with it, planning documents are prepared. These documents present a vision of how the impaired ecosystem is anticipated to appear following completion of restoration tasks and a subsequent period of time while the ecosystem itself recovers the ecological complexity that existed prior to its impairment. There is usually a substantial lag between the time the last task is completed at a restoration project site and the time that the newly restored ecosystem recovers to the point that it corresponds to what was envisioned in planning documents. If the vision was an old-growth forest, it may take one decade to complete all restoration tasks, and several centuries before that vision is fully realized. The time lag is not always this prolonged. Some herbaceous marshes, for example, can be restored to the ecological equivalent of old-growth within a decade using salvaged topsoil techniques, as exemplified in figure 4.1. For most kinds of ecosystems, the waiting time is much longer. This is especially true in terms of fully recovering functionality, as was recently demonstrated by Moreno-Mateos et al. (2012) for a wide range of wetlands.

**FIGURE 4.1.** Recently restored freshwater marsh dominated by pickerelweed (*Pontederia cordata*), Florida, USA.

As an ecosystem undergoes recovery, various properties emerge. These are ecological attributes that will be described at length in chapter 5. They include the resumption of arrested ecological processes and the acceleration of retarded processes; development of community structure and ecological complexity, including habitat and niche differentiation; the capacity for self-organization; the capacity for resilience to disturbance; the capacity for self-sustainability in a dynamic sense; and the capacity for biosphere support. These properties combine to bestow a quality of naturalness to an ecosystem that we strive to recover above all else.

A photograph of an ecosystem is not natural, at least not in the way that the ecosystem is natural. Instead, naturalness is an emergent property of ecological processes as living organisms interact with each other and their environment. These interactions are the interplay of matter within a context of space and time. *Temporal* continuity, is an essential condition for the occurrence of these interactions. That's why a photograph isn't alive. Physical changes, as expressed in terms of form and structure in biotic communities, are inevitable consequences of these interactions. We can state that naturalness is dynamic, because it is the expression of change across time. An ecosystem is characterized by what we call historic continuity and by ecological processes that cause its biotic community to express continual change. Continuity as the consequence of change is a marvelous paradox. This paradox leads us to the most essential principle in ecological restoration; that *we assist in the recovery of historic continuity in a dynamic ecosystem that was temporarily interrupted by impairment.* It is the recovery of this continuity—the essence of naturalness—that brings great satisfaction to those engaged in ecological restoration.

If we walk into the offices of a philanthropic institution to request funding to recover historic continuity and naturalness, they will probably escort us politely to the exit. A more persuasive approach would be to compare our impaired ecosystem to a train wreck and point out that we need funding to put the train (the impaired ecosystem) back "on track," that is, to help it reestablish its temporarily interrupted historic ecological trajectory. A train wreck is a useful metaphor to explain what ecological restorationists do. We used it in the first edition of this book (Clewell and Aronson 2007, 82–83); however, it has a major drawback. A train runs on a track that was laid in advance, and the train can only arrive at a predetermined endpoint, whereas the trajectory of an ecosystem is open ended.

We here suggest a maritime analogy, which avoids that flaw. Recovering historic continuity, we can say, is analogous to rescuing a ship that ran aground on rocky shoals. The ship represents an ecosystem, and the ship's course, as evidenced by its wake, is its trajectory. The shoals represent impairment. As restoration practitioners, we make necessary repairs to recover the leaking hull (i.e., we perform ecological restoration). Thereafter, the ship gets under way and resumes its voyage along its original course, meaning that it is now reconnected to its historic trajectory and has reestablished its historic continuity. We can't predict with certainty the longer-term trajectory of the ship, because it may be blown off course by storms (external environmental conditions), or it may change direction owing to navigation by its officers (internal flux). Nonetheless, we feel confident that the ship will remain in service indefinitely (i.e., it is self-organizing and sustainable), with only periodic need for refueling and service (ecosystem management).

Returning to the definition of ecological restoration (SER 2004), we reiterate that ecological restoration is the process of "assisting" the recovery of an ecosystem. Assistance requires premeditation and intent. Recovery through ecological restoration is therefore an intentional, premeditated process. Ecosystems may undergo natural regeneration, also called self-recovery, spontaneous recovery and unassisted biological succession. There should be no intervention if self-recovery has already done the job. Authors commonly use the term *passive restoration* as a synonym for natural regeneration. Regardless of its popularity, the term passive restoration is an oxymoron. We don't use it in this book and hope that subsequent authors will follow suit. Intervention generally infers some kind of *manipulation* of the biophysical environment at a project site or in its surrounding landscape. Not all interventions involve biophysical manipulation, as we will describe in chapter 8, but they are premeditated, intentional, and therefore not passive.

## Ecological Views of Recovery

It should be obvious that ecological restoration is part assistance by practitioners and part natural regeneration. We will consider the balance between the two in chapter 8. In this chapter, we turn our attention to the principal modes of recovery by natural regeneration, because at some point practitioners will have to determine when the ecosystem has recovered to the point that subsequent recovery can continue without their further assistance. Ecologists generally view ecological recovery from two perspectives. One is *community assembly*, which is concerned with determining what species will participate in the formation of a new (or restored) biotic community. The other is *succession*, which is concerned with how species become rearranged to form an enduring community, sometimes called an endpoint, which can maintain itself in a state of ecological equilibrium. Reduced to their essential elements, community assembly determines what gets established at a site, and succession determines what happens to it subsequently (Hobbs et al. 2007, 163). The separation between assembly and succession is not really that distinct, because species composition, as determined by assembly, can strongly influence succession.

Succession theory has generated an enormous literature over the past century with respect to how orderly or chaotic the process of succession is and the degree of precision to which a specific endpoint can be predicted. Community assembly—a much newer view of recovery—appears to be catching up in terms of interest and research papers that arc generated. We will attempt to sort out important themes in this literature and its attendant terminology in sufficient detail for readers who want to follow discussions on these topics by restoration ecologists. We begin with succession, followed by community assembly. Later, we introduce two additional views of ecological recovery. One is the multiequilibrium concept, which pertains to the potential for contrasting alternative steady-state ecosystems to occupy the same site. The other is the nonequilibrium concept, whereby ecological equilibrium is never attained and the ecosystem remains in a perpetual state of becoming.

### Succession

Succession, also called natural succession, biological succession, or ecological succession, can be defined as the sequence of changes in biodiversity that occur as an ecosystem progresses in its development toward ecological equilibrium.

Classical succession theory originated with Cowles (1899), who described turnover within plant communities that he observed occurring on a sequence of successively older dunes along the shore of Lake Michigan, USA. Clements (1916) expressed the results of Cowles in terms of succession theory, also called *climax* theory. An entire sequence of community development that ends in ecological equilibrium is called a *sere*, and intermediate stages within a sere can each be called seral communities. The endpoint of a sere according to classical succession theory was the climax community, which was presumed to be able to reproduce itself indefinitely in a steady state which, in turn, reflected the now-outmoded concept of the balance of nature (Pimm 1991). Ecological equilibrium is a recent term that has largely replaced the older term of climax community with its narrower and more rigid connotation that now seems misleading.

The facilitation principle underlies much succession theory. The facilitation principle asserts that those plant species that appeared early in a sere will modify and ameliorate the harsh physical conditions of a disturbed or otherwise open site by their shade, accumulations of discarded plant organic matter, and establishment of decomposer organisms in the soil to the point that other plant species, which are better adapted to these improved conditions, will replace them. In other words, entire plant communities replace each other in sequence until the climax community is reached. Community composition as predicted by facilitation can be modified by competition or other inhibitory effects of previously established species that prevent colonization by otherwise expected species. This effect has been called the inhibitory principle.

Succession theory assumes the existence of strong internal regulation and feedback loops, in which one process reinforces the next in sequence. Regulation can be imposed by harsh climatic conditions, such as prolonged below-freezing winter weather in boreal forest and tundra. Plant species composition is limited in such regions, and so is the potential for species composition and community structure. The significance of this observation is that classical succession theory is most applicable in regions where the number of plant species is limited to those that tolerate relatively harsh physical conditions.

Shortly after succession theory was proposed, it was widely accepted to explain ecological development on newly exposed sites (called primary succession) and ecological recovery at disturbed sites (called secondary succession). The theory was alluring on account of its parsimonious simplicity and mechanistic predictability, which allowed ecologists to equate their discipline with the precision of the physical sciences. They asserted that any given region could be characterized by a single climax community, called a monoclimax, toward which all succession eventually led. However, as exceptions and objections emerged, the elegance of succession theory diminished. Gleason (1939) proposed a polyclimax theory, whereby succession could lead to more than one endpoint within a landscape on account of differences in soils and physiography. Other ecologists raised objections relative to the widespread recurrence of disturbances caused by fire or by herbivory from roaming herds of large animals such as bison or elephants. Ecological communities that have been arrested in this fashion were called disclimax (short for disturbance climax) communities.

Owing to these and other exceptions, succession has come to mean the process of vegetation change (Hobbs and Walker 2007) without inferring a directional convergence toward climax or ecological equilibrium. Concepts of succession and climax have thereby lost at least some of their power of prediction, which reduces the status of succession from a theory to a descriptive term that indicates developmental change in biotic expression over time. Some corollaries of succession theory, particularly the facilitation principle, retain their importance.

Succession becomes a much more plausible and useful explanation of ecosystem recovery if viewed from the standpoint of structure rather than species composition. In much of the world, succession theory serves quite nicely to describe the recovery of impaired ecosystems, as long as emphasis is placed on dominant life forms, dominant plant strategies, sizes attained by dominant species, stratification, the relative abundance and frequency distributions of species, and other aspects of community structure, without emphasis on a particular species composition. Ecosystem services are not as dependent on particular species as on community structure and groups of species that are sometimes called guilds (Simberloff and Dayan 1991) and more often, nowadays, functional groups. From this perspective, succession is the process by which species self-organize into sustainable yet adaptive communities. Succession theory is much more palatable—and useful—when the point of reference invoked is an ecologically dynamic community structure with an indefinite and variable species composition, rather than a "climax" community with rigid species composition.

## Community Assembly

Ecologists began exploring the principles—or rules as they are called—for how species assemble to form a new community as an extension of investigations into island biogeography (MacArthur and Wilson 1967). An initial question was to determine which of numerous potential species on the mainland would be successful in dispersing and establishing a new community on an unoccupied island that was only large enough to sustain a few species. Other investigations regarding assembly were made on the behavior of zooplankton communities under laboratory conditions. The search for assembly rules (Diamond 1975; Weiher and Keddy 1999) later expanded to field studies on plant populations, including investigations on the applicability of assembly rules in the design of ecological restoration projects (Young et al. 2001; Temperton et al. 2004).

According to logic, species that initially colonize an open site are more likely to survive and characterize the new community than are subsequently colonizing species. The site could be large such as a newly formed oceanic island or a microsite such as exposed soil deposited at the entrances to burrows by mammals in a grassland. If two identical open sites are simultaneously colonized by the same species but in different sequences, the resulting communities will differ in terms of species survival and abundance (Bastow Wilson et al. 1996). Immigration history affects community structure and ultimately determines ecosystem processes. The principle involved is called the priority effect or historical contingency. Species having "priority" are those that arrived first. The characterization of a new community is contingent on the order and timing of colonization. The priority effect is stochastic (random) and does not lend itself to prediction. That means that any species from the regional pool of potential contributor species could arrive first at a site, assuming equality in the capacity of species to disperse.

Investigations into assembly rules seek to identify environmental filters that control which species from a larger regional pool are allowed membership into a community (i.e., those that are able to become established). Various kinds of filters have been identified including dispersal filters, that is, those factors that allow or prevent a species from having access to a site; tolerance filters (also called ecological filters) that determine which species can become successfully established in spite of harsh physical conditions following dispersal; and disturbance filters that allow some species to survive disturbance events following establishment while excluding or eliminating others (White and Jentsch 2004).

Long before community assembly became a common term in ecology, Frank Egler (1954) noted that the vegetation that first colonized an open site could—under some circumstances—persist in a state of equilibrium, and succession would not occur. This observation presaged the advent of the priority effect concept and caused consternation among ecologists of that day who advocated succession theory. Egler referred to ecosystem assembly of this kind as the phenomenon of initial floristic composition (IFC). Communities tend to develop in response to IFC where stress-tolerant species prevail (Grime 1979). The Florida Everglades illustrate an extensive ecological landscape that developed in accord with the IFC principle and that is dominated by stress-tolerant species. Five thousand years of peat accumulation, formed principally from sawgrass (*Cladium jamaicense*) reveals a remarkably consistent species composition by species that had to cope with low availability of nutrients, fluctuations in oxygen availability, frequent fires, frequent hurricane-driven storms, and occasional salinity shocks.

IFC probably explains ecological recovery over much greater portions of the globe than is currently realized, owing in part to an overemphasis on succession theory as it is taught in most ecology courses. Ecosystems undergoing restoration that are described in the first three virtual field trips (VFT 1. 2, 3) all develop in accord with the IFC model, or any seral development is abbreviated and consists primarily of demographic adjustments in species populations that are already present. In all three of these projects (xeric longleaf pine sandhills, Mediterranean steppe, shola grasslands) it is important to reintroduce any missing plant species that are characteristic of these ecosystems during the course of restoration, and not depend on their spontaneous reintroduction at a presumed later stage of development.

Succession theory is more applicable in environments that favor what Grime (1979) called competition-tolerant species. These environments include great temperate deciduous forest regions of eastern North America, eastern Asia, and much of Europe, where the facilitation principle of succession theory readily explains ecological recovery. Succession theory developed in universities located in these regions by investigators who were influenced by their local environment and who tried to impose succession theory on large regions of the world where it was not appropriate. The message of the IFC phenomenon to restorationists is that species initially established at a project site may persist indefinitely. Species and *provenance* selection can have long-lasting ecological consequences and require careful consideration and planning.

Although historical contingency is germane to any discussion of assembly rules, so are the antecedent conditions of the site that is being colonized. These conditions include the usual suite of environmental factors by which a site is characterized, such as soil type, hydrological input, climate in terrestrial locations and substrate, salinity, temperature, and currents in aquatic locations. Antecedent conditions also include ecological legacy, which consists of all of the live and dead organic matter that survives impairment and contributes to the assembly of a new community. For example, fallen trees serve as nurse logs that provide microhabitat for colonizing seedlings and thereby contribute to habitat heterogeneity (Maser and Sedell 1994). Surviving physical structures contribute to ecological legacy, including pits and mounds of exposed mineral soil that form when trees fall (White and Jentsch 2004). Ecological theorists interested in assembly rules refer to the effects of antecedent conditions by the term ecological determinism. Determinism connotes predictability in the community assembly process, whereas the priority effect is stochastic and unpredictable. Community assembly is governed by a balance between the priority effect (historical contingency) and determinism.

Research in community assembly is also less concerned with particular species than it is with functional groups consisting of similar species that perform the same ecological process, or that respond in similar ways to a given disturbance. Native species within a community or community type that belong to the same functional group are assumed to be interchangeable in respect to performing ecological processes, as long as both the environment and the community are stable. Ecosystems are dynamic; consequently, species interchangeability is more easily conceived in theoretical constructs than can be demonstrated in Nature.

The assignment of a species to a particular functional group is made easily in a research setting when only a single process is under consideration. However, a given species generally participates in more than one ecological process and may retain membership in multiple functional groups. Assignment of functional groups becomes more complicated when planning an ecological restoration project in which all ecological processes must be considered holistically. Nonetheless, the conceptualization of functional groups is a useful exercise in the development of reference models and restoration project design, and we will return to their consideration in chapter 5.

## Multiple Equilibria

In recent years, ecologists recognized that more than one potential endpoint or ecological steady state may develop at a single site, depending on variations in ecological development following disturbance. This insight corresponds to the multiequilibrium concept (also known as metastability), as summarized by Hobbs and Suding (2007). The multiequilibrium concept pertains to landscapes where progressive degradation causes the sudden replacement of one ecosystem by another that is generally inferior in terms of biodiversity and ecosystem services. Ecologists have explained community replacement of this type in terms of state-transition models (sometimes written as state and transition). The state-transition concept describes abrupt switches between alternate ecosystem states following an irreversible change in environmental conditions. The environmental tipping point between alternative states is called a threshold of no return, or "threshold of irreversibility" described in chapter 2.

State-transition models were first developed by rangeland ecologists (e.g., Westoby et al. 1989; Milton et al. 1994), who identified exogenous disturbances as forcing factors instead of priority effects to explain alternative states caused by land use and land management. State-transition theory has demonstrated the need for ecological restoration to recover former states but has not elucidated new tactics for practitioners to switch an affected ecosystem back to its prior state. Grant (2006) applied state-transition notions to detect and evaluate developmental stages at a restoration project site, but no thresholds were crossed and no switches to alternative states occurred. The potential for alternate stable states increases the opportunity for restorationists to select a suitable target for a restoration project. In chapter 6, we describe examples of alternate stable states and discuss the selection of appropriate targets for restoration.

### Nonequilibria

According to the nonequilibrium concept, many ecosystems develop and evolve but never reach equilibrium on account of continuing natural flux in species diversity from patch dynamics. Ecologists apply the term *patch dynamics* to describe the ecological effects of localized stress and disturbance events that occur in small areas (microsites) within a larger community or landscape (Pickett and White 1985). According to this concept, localized succession within microsites leads to a perpetual state of nonequilibrium, owing to the pervasiveness of stochastic outcomes of internal ecological processes combined with external stress events that add to the unpredictability of ecological expression. Any biotic community is therefore constantly evolving in a perpetual state of ecological flux, as Westoby et al. (1989) and Wessels et al. (2007) have described for natural rangelands.

What appears as a generalized community covering a large landscape actually consists of populations occurring in a mosaic of small patches, each in its own phase of development in response to local biophysical conditions. This perspective of an ecosystem consisting of different-aged microsites has been called the "carousel model" by van der Maarel and Sykes (1993). Palmer et al. (1997) use the term "lottery rules" to refer to the stochastic nature of the carousel model. These analogies emphasize small-scale stochastic turnover wherein myriad unpredictable recruitment or disturbance events can take place at a fine scale of resolution. Fine-scale turnover ensures that species and community structures characteristic of all developmental stages of a recovering ecosystem are represented and persist indefinitely. At least a few individuals of the most competition-sensitive, r-strategist species remain extant somewhere in a mature ecosystem and will skip about, so to speak, from one microsite to the next as sites become available for colonization.

Patch dynamics may be a much more common phenomenon than many realize. When we walk through a forest, our eyes tend to see the largest trees, and we ignore—or look right through—canopy gaps without registering the seral dynamics that are occurring before our eyes. Those gaps form when a large tree falls, thereby creating localized succession in a microsite where intense competition is occurring to determine which one of many saplings will eventually replace the fallen tree. Likewise, anyone who has sampled plants in a grassland or savanna probably has been surprised at the irregular dispersion of many species and their shifting patterns of dominance, despite the apparent uniformity of the environment as one casually walks through it. These irregularities are commonly attributable to patch dynamics.

Some restoration ecologists have embraced nonequilibrium theory. Hobbs and Norton (1996) asserted that disturbance, rather than stability, is the normal state in an ecosystem. They explained that the long-held concept of equilibrium has given way to a nonequilibrium concept, in which ecosystems are characterized more by instability than by permanence. Flux and resulting patch dynamics are recognized as the primary characteristics of the nonequilibrium concept (Suding and Gross 2006), which stand in direct contrast to the single equilibrium concept (or monoclimax) embraced by classic natural succession theorists. The nonequilibrium concept posits that a stable state or steady state cannot be reached as predicted by succession theory because ecosystems undergo continuous flux. The *SER Primer* (SER 2004, 1) arrived at this same position by stating that "the restored ecosystem will not necessarily recover to its former state, since contemporary constraints and conditions may cause it to develop along an altered trajectory." The nonequilibrium concept largely explains why ecosystems and ecological restoration are openended and why targets selected at the start of restoration projects are not necessarily attained when the restored ecosystem achieves ecological maturity.

## Ecological Theory and Restoration

Ecological theory poses practical challenges for those engaged in the restoration of ecosystems. For example, to what degree, if any, should a restoration project be designed to rely on the priority effect (historical contingency) for populating a restoration project site with species? The other option is to introduce species intentionally. For the restoration of estuaries, only a few species may need to be intentionally introduced to provide initial structure, such as corals, oysters, and seagrasses. There would be no need to introduce planktonic algae, and widely dispersed zoospores of macroscopic algae will colonize if the substrate is favorable. Spontaneous dispersal, which could activate a priority effect, determines the rest of the estuarine community.

In terrestrial communities, desirable species of plants are commonly introduced intentionally; particularly if these species are not likely to populate the restoration site because of lack of proximity to the nearest seed source or some other factor. The balance between reliance on natural regeneration (priority effect) and intentional introductions (ecological determinism, also socioeconomic pragmatism) falls squarely on the desks of those who design restoration projects. There is no avoidance, because an option for natural regeneration requires a conscious decision. We cannot avoid making decisions in ecological restoration.

Another choice to be made is how much opportunity for flux and resultant nonequilibrium should be incorporated into restoration plans? This question is particularly germane to semicultural ecosystems, where the human element enters substantially into the equation. For example, in Europe, there is much activity in restoring beloved chalk meadows that once graced large rural areas and were the destinations of hikers and picnickers and the settings of choice for poets and novelists (Willems 2001). These meadows harbor large numbers of native plant species, including some that are rare and red-listed for protection by the International Union for Conservation of Nature (IUCN). They are also early seral communities that disappear quickly if overtaken by forest. To preserve their biodiversity, they are intentionally mowed each year, sometimes with handheld scythes in the traditional manner. Animal manure is sometimes spread. If the goal of ecological restoration were a forest, then a forest could be restored at the project site directly without an intervening chalk meadow stage. Forest and chalk meadow are two contrasting alternative ecosystems. Which one is to be restored is a decision for stakeholders and project sponsors (chap. 9).

Yet another consideration before those who design ecological restoration projects and programs is that of accelerated recovery. Under natural conditions, seral development often occurs over protracted periods of time, and Nature has no inherent impetus to rush. Ecological restoration affords the opportunity to accelerate recovery by introducing species typical of later stages of development during project implementation and adjusting the abiotic environment as needed to favor their persistence. Although individuals of these species that are initially introduced at a project site may persist, they probably will not proliferate, at least not right away, until their habitats differentiate. For example, organic matter may have to accumulate in the soil to support large populations of mycorrhizal symbionts before late-successional trees can flourish. One restoration tactic to accelerate succession would be to amend the soil in places with organic matter, inoculate it with mycorrhizal fungi, and outplant nursery stock of a late-seral tree species. After sufficient passage of time for the restored ecosystem to develop suitable habitat, seeds from these planted trees will be available onsite and expedite the proliferation of that species. If this tactic had not been included in the restoration design, natural dispersal of that tree species from external sources may take much longer to occur, or never occur at all, depending on propagule availability offsite.

Many ecological restoration projects are intended to supplant assembly processes and succession. Instead, biophysical manipulations are performed to recover an impaired ecosystem sufficiently for its ecological processes to resume at normal levels without the need for that ecosystem to undergo a sometimes lengthy natural process of recovery. Again, a medical model can be invoked, whereby the physician uses pharmaceuticals, surgical procedures, and physical therapies to obviate the need for a patient to wait for an indefinitely long period of time before natural healing occurs.

Restoration practitioners need to be familiar with concepts of community assembly, succession, and other modes of understanding recovery under natural conditions, so that impaired ecosystems can be nurtured and their recovery expedited in an ecologically sound and cost-effective manner. To reiterate, ecological recovery is a deceptively simple term for a surprisingly complex concept.

# Chapter 5

# Ecological Attributes of Restored Ecosystems

In this chapter we identify and describe eleven ecological attributes that characterize ecosystems as being successfully restored. The reappearance of these attributes signifies that ecological recovery as described in chapter 4 has occurred satisfactorily. The first four are directly attainable attributes that manifest in response to biophysical interventions conducted by practitioners at restoration project sites. These attributes include an appropriate species composition as determined by the reference model, initial development of community structure, an abiotic environment that supports the biota, and a landscape context that facilitates normal flows and exchanges organisms and materials with surrounding areas and that lacks threats to the restored ecosystem.

The other seven ecological attributes are indirectly attainable, meaning that they appear or emerge on account of interactions of organisms with each other and their abiotic environment and not because of manipulations conducted by practitioners at a project site. These indirect attributes are presumably realized if interventions by practitioners were performed satisfactorily for the attainment of direct attributes. Indirect attributes include the reestablishment of ecological functionality in terms of ecological processes, the reestablishment of historic continuity, development of ecological complexity and the capacities for self-organization, resilience, self-sustainability, and biosphere support. Unlike the four direct attributes, the indirect attributes are not readily measured and documented. In most ecological restoration projects, some of these attributes can only be partially satisfied, if at all, for unavoidable technical and pragmatic reasons. The important point is that everyone involved in a restoration project seriously explores all available avenues for their fulfillment before the quest for any of them is abandoned.

Any ecosystem, whether or not it has ever been restored, can be said to exist in a state or condition of wholeness if it possesses these eleven attributes. The term *holistic ecological restoration*, as used by Clewell and Aronson (2007), refers to restoration work that is intended to return an impaired ecosystem to wholeness in regard to these eleven attributes, as informed by a well prepared reference model.

Aronson et al. (1993a) initially advocated the restoration of ecological attributes. Clewell (2000a) prepared a more generalized list of attributes, which were later condensed and published in the *SER Primer* (SER 2004, 3–4). In this chapter, we combine some attributes from the *SER Primer*, add some others, and refine some descriptions (table 5.1). The attributes are portrayed diagrammatically in figure 5.1 to indicate how one influences another as an ecosystem recovers.

## Species Composition

Plant species composition represents the basic trophic level of an ecosystem (the producers), and plant species composition influences and ultimately governs all other ecosystem attributes (table 5.1). Ensuring a comprehensive and appropriate plant species composition is, therefore, the principal obligation of restorationists in both terrestrial and aquatic systems, even where phytoplankton prevails. Ideally, plant species composition consists of those species that occurred prior to impairment in the intact ecosystem, with allowance for substitutions of species from the same guild or functional group, where necessary. However, better adapted species may be required at project sites where environmental conditions have changed. Justification for substitutions of species assumes that the regional species pool for a community is larger than the species composition at a particular locality, such as a restoration project site. Approximately the same number of plant species should occur in the restored ecosystem as occurred in the preimpairment ecosystem. Underrepresentation of species could lead to inefficient functionality and instability. More species could be introduced in order to induce competition that favored the fittest species. We use the term "comprehensive species composition" to indicate one that closely approximates that which occurred prior to impairment, with substitutions that are ecologically justified.

Aldo Leopold (1887–1948) wrote passionately about the importance of species composition:

> The last word in ignorance is the man who says of an animal or plant: "What good is it?" If the land mechanism as a whole is good, then every part is good, whether we understand it or not. If the biota, in the course of eons, has built something we like but do not understand, then who but a fool would discard seemingly useless parts? (Leopold 1993, 146–47)

TABLE 5.1.

*Ecological attributes of restored ecosystems*

**Directly attained attributes**

Species composition: The restored ecosystem contains a comprehensive assemblage of potentially coadapted species as informed by the reference model. The species include representatives of all known functional groups. The species are indigenous, and invasive organisms are absent insofar as possible.

Community structure: Species populations are established in sufficient abundance and distributed across the project site adequately to facilitate structural development in the biotic community.

Abiotic environment: The abiotic environment has the physical capacity to sustain the biota of the restored ecosystem.

Landscape context: The restored ecosystem is suitably integrated into a larger ecological matrix or landscape with which it interacts through abiotic and biotic flows and exchanges as informed by the reference model. Potential threats to the health and integrity of the restored ecosystem from the surrounding landscape are eliminated insofar as possible.

**Indirectly attained attributes**

Ecological functionality: Ecological processes in the restored ecosystem occur normally for its ecological stage of development, and signs of dysfunction are absent.

Historic continuity: Biodiversity recovers to the point that the historic ecological trajectory of the ecosystem, which was interrupted by impairment, is reestablished.

Ecological complexity: The ecosystem develops complex ecological structure that facilitates niche differentiation and habitat diversity.

Self-organization: The ecosystem develops feedback loops that increase its capacity to conserve its resources and increase its potential for autonomy.

Resilience: The restored ecosystem is sufficiently resilient to resist or self-recover from all but the most severe disturbance events and to benefit from stress events that maintain ecosystem integrity.

Self-sustainability: The restored ecosystem is self-sustaining to the same degree as its reference ecosystems and has the potential to persist indefinitely. Aspects of its biodiversity may fluctuate or change in response to internal flux and external environmental changes.

Biosphere support: The restored ecosystem generates atmospheric oxygen, absorbs $CO_2$, facilitates thermal reflectance, and provides habitat for rare species.

Leopold famously continued with the analogy of repairing an old fashioned pocket watch, whose mechanism relied on moving parts rather than a digital display: "To keep every cog and wheel is the first precaution of intelligent tinkering."

Leopold's message is clear. When we restore an impaired ecosystem, we must ensure that all of its parts from the preimpairment state have been accounted for.

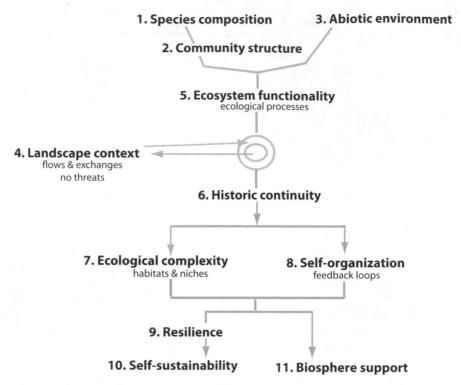

**FIGURE 5.1.** Eleven ecological attributes, their principal relationships, and their approximate order of appearance (allowing for considerable overlap) as an ecosystem develops during and after ecological restoration.

Of course this is analogy cannot be fulfilled with clockwork precision when recovery is accomplished by living organisms rather than practitioners. However, we can ensure that all of the species are represented that are specified in the reference model. Otherwise, the restored ecosystem may not regain its former structure and may not function as well as previously.

At some project sites, the plant species composition may be unimpaired. At other sites, some species that were eliminated may recolonize the project site by means of natural dispersal and will not require reintroduction. Practitioners should determine which species are already present at a site and which, if any, require reintroduction. Not every appropriate plant species must be present at the time restoration tasks are completed, as long as their natural colonization can be reasonably expected through unassisted regeneration at a later time.

Animal species generally do not require intentional introduction on account of their mobility and their propensity to discover and occupy favorable habitat. In VFT 3 (shola grassland restoration) the removal of trees led to increased areas of surface water to which many animals positively responded—from insects and fish to birds and large mammals. Animals may not come spontaneously to restored habitats that have become isolated by *fragmentation*. Captive breeding, translocations, and reintroductions of animals are therefore sometimes necessary to conserve "all the parts." This has been attempted to various extents around the world, although arguably more for rare species ex situ conservation and the reestablishment of spectacular, wide-ranging wildlife (e.g., mountain sheep) than for in situ ecological restoration. Animal reintroductions may be more important than reconstituting the plant community in instances of restoration where landscape fragmentation and extreme isolation occur. Sometime relatively modest interventions can accelerate the recovery of animal habitat, as is shown by intentional log piles in figure 5.2.

**Figure 5.2.** Log pile as habitat to attract lizards and other small animals at a newly planted jarrah forest on mined land in Western Australia.

Wildlife restoration has become a specialty within ecological restoration, which integrates the principles and procedures of wildlife management with ecological restoration (Morrison 2010). Much of the emphasis is to ensure that project design provides suitable habitat in terms of food and water resources, cover, and territorial space for animals typical of the ecosystem or landscape to be restored. The reintroduction of top predators in particular can be a crucial step in ecological restoration, as we discussed in chapter 3 with regard to trophic cascades.

We continue by addressing several topics that are particularly relevant to species composition in relation to ecological restoration. They are coadapted species, functional groups, redundant species, and alien species.

## Coadapted Species

We become used to seeing many of the same species in the same kind of biotic community as it occurs repeatedly throughout a biome or ecoregion. Ecologists in past generations speculated that certain species combinations endowed a community with a competitive advantage. According to the theory of diffuse *coevolution*, species belonging to well-established assemblages are genetically adapted in ways that benefit their collective survival, fecundity, and capacity for self-organization.

Diffuse coevolution is a concept supported by several lines of evidence (Fox 1988; Inouye and Stinchcombe 2001; Strauss et al. 2005). Evidence from the study of interactions between plants and herbivores suggests that diffuse coevolution involves the development of feedback loops that tightly regulate demographic properties and trophic interactions, which bestow a selective advantage to an entire community. The phenomenon is an extension of the evolution of species pairs, for example, species participating in animal-mediated pollination and seed dispersal, nitrogen fixation, mycorrhizae, and mimicry (Soulé et al. 2005), to the evolution of larger groups of species.

A supplementary or alternative hypothesis to that of diffuse coevolution, which could possibly explain coadaptation of species in a community, is morphic resonance, whereby the forms of previous systems (communities) influence the development of subsequent similar systems (Sheldrake 1981, 95; Sheldrake 2012, 99–101). Morphic resonance draws upon field theory and quantum mechanics and explains phenomena in such terms as basins of attraction. The strength of morphic resonance as a comprehensive theory to explain phenomena in developmental and behavioral biology suggests that it could also explain species coadaptations in communities. We propose morphic resonance as a promising field of investigation for restoration ecologists, although we are aware that it represents a paradigm that investigators may be reluctant to explore (Sheldrake 2012, 229).

When we specify that seeds are to be collected locally to ensure the reintroductions of local *ecotypes* at a project site, we may also be ensuring the reassembly of coadapted species. When we prepare a reference model for a restoration project, we record potentially coadapted species composition with the intent of reassembling them at restoration sites. We may never know if we are reassembling genetically coadapted species that diffusely coevolved under each others' influence, or if we are assembling a community of species that happen to interact in a complementarily favorable way for some other reason. In any event, we will have more confidence that the restored ecosystem will be successful than if we haphazardly chose our species as if from a seed catalog, or "off the shelf" at a commercial plant nursery.

The following hypothetical case illustrates one way coadaptation might work. Assume that twenty-one species of plants comprise the K-strategists in a particular community. The roots of these plants extract water and mineral nutrients from seven different depths. Three species produce roots at each depth. The roots of the three species are metabolically active at different seasons. Therefore, competition for moisture and nutrients is minimized among the twenty-one species spatially and temporally, and they share ecosystem resources equally. Now suppose that this community suffered impairment, and it was restored using twenty-one species ordered from a seed catalog. Further suppose that none of the species listed in that catalog were mutually coadapted or otherwise complementary. By chance, the roots of all the species extracted moisture and minerals at the same depth and in the same season. After a few years, only the fittest species survived, because the other twenty species perished from competitive exclusion. This is obviously an ecological morality tale and not an example from Nature, but it gets the point across: we rely on a well-prepared reference model, because it tells us what species are probably coadapted. Reliance on presumed coadapted species gives us our best shot at successful restoration. In addition, it guarantees historic continuity.

## Functional Groups

A *functional group* consists of those species that perform a particular role in an ecosystem or respond to a given stressor or driver in an adaptive manner (Lavorel et al. 1997; Gondard et al. 2003; Rosenfeld 2002). Examples of functional roles are carbon fixation through photosynthesis; nitrogen sequestration; and decomposition of dead plant materials into humus, herbivory, soil stabilization, and microclimatic control. These are general functional roles. Each of them can be further refined into a series of roles. For example, the reduction of dead plant material may begin with fungi that dissolve pectin that cements plant cells together, continue with insect larvae that digest partially separated chunks of wood, commensals in insect guts that digest cellulose, and bacteria that continue the digestion process in the fecal discharges from insects. Additional functional roles can be identified, such as the further digesting decayed wood that was ingested by an annelid (earthworm) and by species of fungi that specialize in the dissolution of different compounds that comprise wood. A fungus, in turn, may enter into mycorrhizal relationships with the roots of trees with which it exchanges mineral nutrients for carbohydrates. One species can play multiple functional roles. Furthermore, it appears that functional groups, as well as species composition, play a role in maintaining resilience within ecosystems and at the landscape scale as well (Sundstrom et al. 2012). This is obvious for top predators that affect herbivores, and hence plants, in several ecosystems at once (e.g., Schmitz et al. 2004). If key species or functional groups among plants are absent at a restoration project site, they may have to be introduced by practitioners. For example, earthworms may have to be introduced into the new substrate in a physically reclaimed surface mine. With regard to animals, benthic invertebrates and fish can be introduced to restock restored streams and wetlands.

Ecological restoration projects that are conducted in urban and peri-urban areas may preclude the recovery or reintroduction all of their former functional groups, owing to contemporary conditions and constraints. For example, the proximity of residences may prove to be a deterrent to the reintroduction of top predators. Hunters and farmers may have serious objections as well, as happened regarding proposals to return wolves into Oregon and northern California, and many other places. Consequently, populations of herbivores such as deer commonly increase to the point that they threaten newly restored ecosystems with overgrazing and comprise a nuisance to residents whose ornamental plantings are damaged. In such circumstances, practitioners are not likely to be able to introduce top predators to control deer, but they can prepare plans for professionals who will assume management responsibilities of the restored ecosystem. The plans would identify the problem in terms of potential losses of biodiversity and inhibited ecological functionality unless deer populations were culled.

Until recently, functional groups among plants were thought to correspond in good measure to what plant ecologists call *life forms*. For example, all needle-leafed coniferous trees share the same life form in a pine-hardwood forest. However, it is overly simplistic to assume that a given life form corresponds to a specific functional group. For instance, some legume trees fix atmospheric nitrogen in symbiosis with rhizobacteria in their roots; many other legume trees do not. In a given tropical forest, for example in the Amazon, some legume trees of both types are found in the understory, canopy, and emergent strata. Therefore just as canopy trees may not constitute a single functional group, so legume trees of the canopy stratum do not constitute a useful functional group for this forest; instead we'd have to identify all *nitrogen-fixing* legume trees of the canopy stratum. Further, recent studies show that even within clearly coherent functional groups much variation exists in ecophysiological and phenological features. Currently, many functional and behavioral ecologists concentrate on identifying and studying *functional traits* of individual taxa and do not assume any a priori functional correspondence between life forms and functional traits. An example is the suite of fire-related plant traits found in dozens of woody species occurring in the five regions of Mediterranean type climate (Keeley et al. 2012).

Among animals, one key trait with multiple effects on ecosystem functioning is herbivore body size, because it determines the kind and amount of resources individuals among predator populations can exploit (Schmitz et al. 2004). It also influences the herbivores' vulnerability to predation: the more herbivores there are, the more pressure there is on vegetation. Obviously, herbivores vary greatly in size; hence there cannot be a single herbivore functional group. This clarification is especially important in effectively combating the arrival of harmful alien organisms, as we will discuss in the section on that topic.

## Redundant Species

Walker (1992) was the first to note that many ecosystems appear to contain far more species than are actually necessary to carry out particular ecological processes. As the number of species in an ecosystem increases, so does the likelihood for functional redundancy, defined as two or more species performing exactly the same functional role in ecosystem processes. The assumption is that only one species is needed to perform that role, and redundant species are irrelevant, superfluous, and can be ignored in the design of ecological restoration projects. Another assumption is that ecologists know the roles that a given species plays in an ecosystem, and that redundant species can be identified. Neither assumption can be readily confirmed with confidence in the field. Furthermore, we have already established that one species may perform multiple functional roles, and redundancy for one role may be unavoidable for the performance of another role. Redundancy may be essential for other reasons. Naeem and Li (1997) and Wellnitz and Poff (2001) suggested that the capacity of a given species to perform a functional role within an ecosystem varies over time and space, particularly in a heterogeneous and changing environment. Therefore, the presences of several redundant species would ensure that a functional role would be performed satisfactorily, regardless of a variable environment.

Another benefit of precaution that Walker (1992) did not consider is that a species that fulfills a particular ecological role in an intact ecosystem may not do so in an impaired ecosystem. For that reason, seemingly redundant species may be needed to assure proper functioning during ecosystem recovery. Finally, the capacity of an intact ecosystem to evolve in response to environmental change may depend on redundant species that are suited for the new environmental conditions. In other words, an ecosystem can be preadapted to new and unprecedented environmental conditions by carrying redundant species that may function effectively in an altered environment (Naeem 1998). Practitioners facilitate that eventuality by introducing a comprehensive complement of plant species. The precautionary approach in ecological restoration would be to introduce as many of the original species as possible regardless of redundancy.

## Alien Species

Alien species are unlikely to represent the historic trajectory or to be coadapted, although some are known to function in a complementary manner with native species. Invasive alien species preempt substantial ecosystem resources that would otherwise be available for native species. An influx of numerous alien species, even if they are not invasive, may threaten the reestablishment of historical continuity and dilute the ecological benefits of coadaptation by native species. For these reasons, alien species (including subspecific alien taxa) are candidates for removal from a restoration project site, particularly if they are likely to persist and strongly inhibit, or in the case of animals—graze or prey upon native species. The *SER Primer* (SER 2004) advises that restored ecosystems should consist of indigenous species to the greatest practicable extent. The onus remains on project sponsors, planners, practitioners, and stakeholders to decide what is tolerable and practicable. The recognition, disposition, and treatment of alien species in the design and implementation of ecological restoration projects can only be determined on a case-by-case basis with liberal doses of common sense, based on local knowledge of natural history and tempered by pragmatism.

Alien species can be safely planted as cover crops or nurse crops, as long as they are relatively shortlived and are unlikely to reproduce and inhibit establishment of native species. A cover crop temporarily stabilizes exposed soil, and a nurse crop facilitates the establishment of desirable, persistent species. The European annual ryegrass (*Lolium perenne*), for example, is commonly planted at North American restoration sites as a cover crop. Nurse crops are planted to increase the organic matter content of mineral soil or, by planting legumes, to increase the nitrogen in the soil, although that practice can stimulate the growth of competitive weeds. Fast-growing and short-lived alien trees are sometimes grown as nurse crops that serve as wind breaks and shade to protect recently established forest species from lethal full exposure to solar radiation (Román-Dañobeytia et al. 2011). Planting nonnative species for such purposes requires knowing that the benefit far outweighs the risk of invasion.

Some noninvasive alien species can also be ignored by restorationists, if they were introduced into a region quite some time ago and appear to have become integrated into the local community as participants in its ecological processes. These are sometimes known as naturalized species; however, that term is not uniformly applied. Naturalization of alien species is particularly common in semicultural ecosystems. Examples in the tropics are naturalized fruit-bearing trees such as mango and papaya. Many alien weedy species coevolved concurrently with the domestication of certain crop plants under conditions of cultivation that favored both species. These so-called segetal species would be appropriate to include or accept in the restoration of a semicultural ecosystem of the kind where they evolved.

Careful thought needs to be given to the costs and benefits of removing specific alien species. We are aware of attempts to remove nonnative species from wetland restoration projects sites by labor crews that trampled and destroyed many desirable plants while mechanically removing naturalized species. In other sites, nonnatives were treated with broad-spectrum herbicides that killed neighboring plants of desirable native species with which they grew. In these instances, public agencies with regulatory powers demanded eradication of naturalized species just because they were nonnative. Some of these species cannot possibly be eradicated and would undoubtedly reestablish themselves after eradication efforts were no longer required upon completion of all other project work. It would have been better to ignore these naturalized species as targets for eradication. Blanket mandates of this sort are overreactions to satisfy agency policy rather than to facilitate ecological restoration. Regulatory personnel are quick to develop stipulations in environmental permits to remove or control alien species, perhaps because the presence of alien species is more easily monitored than are most other criteria for determining successful ecological restoration. The use of alien species issues as surrogates for rigorous assessment of restoration prevents the application of a more rational approach to alien species issues.

## Community Structure

By community structure, we refer to the shape or physiognomy of the community or, if you will, its architecture. Community structure is three dimensional and can be described in terms of its vertical and horizontal components. Although structure ultimately reflects the species that comprise it, we distinguish community structure from species composition in this book. Biotic communities have their own characteristic arrangements of organisms, which are determined largely by their abundance and juxtaposition and the degree of homogeneity or heterogeneity of species populations. Structural complexity offers surfaces where ecological processes occur. The more structure that is available, the more opportunity there is for organisms to interact.

For example, submerged surfaces of wetland plants are commonly coated by sessile diatoms and filamentous green algae that collectively comprise what is called periphyton. In turn, the periphyton contributes much of the primary production and forms the basis of food chains in these systems. Without structure provided by emergent wetland plants, periphyton disappears. Plant life form contributes both structure and texture. Vertical stratification of the vegetation may contribute more structure than the spatial distribution of species. Sedentary benthos may contribute much structure, such as oysters. The abiotic environment offers additional structure, such as rock outcrops and other topographic unconformities, which, in combination with biotic structure, add even more structural diversity to overall ecosystem structure. Large woody debris (as shown in figure 5.3 and described in VFT 6), and other dead organic matter, even humus, also contribute structure in an ecosystem.

Project tasks in ecological restoration generally allow practitioners to initiate

**FIGURE 5.3.** Large woody debris used in stream restoration both as armament to redirect flows and stabilize banks, and as habitat for macroinvertebrates and fish, in the Olympic Peninsula of Washington state, USA.

the development of community structure by introducing desirable species and removing undesirable species. Both the spacing of outplanted nursery stock and how evenly a species is planted across a project site contribute to the eventual development of community structure. Project plans commonly determine spacing and evenness, and project planners need to be aware of the influence of planting designs on the development of community structure. For instance, VFT 1 describes how pine trees were managed during restoration tasks, so that they contributed flammable leaf litter as fuel for surface fires, and response to fire was the primary determinant of community structure. At most restoration project sites, restoration practice can only account for a small proportion of the development of community structure. Most structure develops after project work has been completed by means of continued growth and demographic processes. Restoration tasks usually nudge the development of structure in a favorable direction. Subsequent development of structure is described under the heading of ecological complexity.

## Abiotic Environment

The abiotic environment, which comprises the physical infrastructure of a restored ecosystem, should be capable of supporting the evolving biota as an ecosystem recovers. Otherwise, the desired species composition and community structure cannot be sustained. Impairment of the physical environment in terrestrial and wetland systems largely concerns alterations to hydrology, water quality, and soils. In coastal environments, salinity becomes an equal concern. Ditching and drainage—or impoundment and flooding—may alter the water table, the seasonal availability of moisture, and the amount of soil saturation or inundation and other aspects of hydrology. Impacts on hydrology commonly reduce the residence time of water in catchments, increase peak discharges in streams, and prolong intervening periods of low discharge. All such changes can degrade ecosystems. Plants and animals become stressed by altered schedules of moisture availability. If soils become desiccated, their organic matter can oxidize and cause profound degradation. If soils become saturated or inundated more than briefly, they may become anoxic and unable to support organisms that require aerated soils. Soil can suffer from other impacts, such as erosion and the compaction and mechanical damage from overgrazing by livestock or trafficking by heavy equipment.

At restoration project sites on former agricultural lands, soils sometimes retain excessive amounts of nutrients and altered pH that attract highly competitive r-strategists and invasive species. At such sites, the soil is sometimes amended with sawdust or woodchips, so that fungi that decompose cellulose will metabolize nitrogenous compounds and thereby reduce fertility. At other sites, the topsoil and its high nutrient content has been scraped away, as was described in VFT 2 for Mediterranean steppe restoration, effectively removing what could be considered pollutants.

Degradation and damage to aquatic systems include hydrological alteration that affects water quantity and seasonal changes in water volume, such as stream discharge; alterations to water quality, such as increased turbidity; increased pollution; altered water chemistry and temperature; or a change to the substrate, such as scouring, dredging, or sedimentation. The cause of degradation in aquatic systems is commonly ascribed to impacts occurring in terrestrial systems in the catchment. Timber removal, for example, can cause accelerated surface runoff, pulsed discharge into receiving waters, elevated water temperature, increased turbidity, reductions of inputs of detritus upon which benthic invertebrates feed, and the deposition of eroded sediments in stream channels. Discharge of animal wastes and fertilizers in agricultural runoff is a common cause of accelerated eutrophication. In estuaries, aquatic systems can be degraded or damaged from changes in salinity which, in turn, can be caused by water usage inland that alters the volume or seasonality of river discharge.

Issues of hydrological alteration can be approached either by repairing the physical environment, so that the preimpairment *hydroperiod* returns, or by substituting another kind of ecosystem that is suited to a new hydroperiod. The latter may be the only option, when, for example, an impoundment or other public works projects have permanently altered the hydrology of an ecosystem. In such instances, impoundment would constitute an irreversible contemporary condition that crosses a threshold of no return.

Repair of the physical environment must be accomplished very thoroughly and carefully. Otherwise, subsequent manipulations to the biota, such as the intentional reestablishment of vegetation, could fail. Retrofitting the physical environment at that stage of project work can prolong a project's duration and inflate project costs to an amount that far exceeds the amount that would have been needed to do the job correctly in the first place. The ultimate test of the suitability of the physical environment is its capacity to sustain reproductive species populations of an appropriate biota for a self-sustaining ecosystem of a designated type. In other words, the biota serves as a bioassay for the physical environment.

## Landscape Context

Organisms, energy, water, nutrients, and other matter move freely between ecosystems, across landscapes and sometimes continents. Ecosystems depend on these flows and exchanges for their ecological functioning. When an ecosystem is restored, an essential task is to ensure that these flows and exchanges occur at what are assumed to be normal levels. Migrant birds provide obvious examples. A blackburnian warbler (*Setophaga fusca*) may nest in Canada and overwinter in Ecuador, while transporting water, nutrients, and energy wherever it goes in its annual migration.

During the design phase, the landscape context of a project site requires consideration. If the landscape is impaired, the restoration effort may not attain its full potential. Consider for a moment that the landscape surrounding a restoration project site had been cleared of its vegetation for some permanent land use. Animals may not be able to move in and out of the project site, because the cleared landscape makes them susceptible to predation. In other words, the quality of a restoration project depends in part on the quality of the larger landscape in which it is set. Restorationists need to integrate the landscape with the project site in as much as possible. If there are deficiencies in the landscape context—existing and potential—stakeholders and project sponsors should be aware of them prior to making a decision to implement a restoration project.

Threats in the surrounding landscape that may affect the quality of a project should be addressed by practitioners insofar as possible (see VFT 3). For example, surface runoff from cleared landscape could move rapidly onto the project site, causing erosion or depositing sediment. If nearby colonies of invasive plant species on this cleared land pose a concern of invasion, the practitioner could negotiate with adjacent property owners or managers to eradicate them as one of the restoration tasks. The cost of doing so would likely be much less than treating invasive species onsite that repeatedly colonized from the surrounding landscape. Figure 5.4 illustrates another instance where stream erosion threatens woodland restoration. One of us was engaged with an issue of this sort on mined land where

**Figure 5.4.** Tree planting for woodland restoration in New South Wales, Australia. The longer-term integrity of a restored woodland is threatened by the eroding stream channel, which could be restored concurrently by installing sills at intervals. Sills would trap sediments and raise the elevation of the stream bottom.

lowland forest was being restored (Clewell et al. 2000). Seeds produced by exotic pasture grasses that had been planted upslope were carried by surface runoff after every rainfall event into the project site, where herbicides applications could not keep pace with waves of germinating grass seeds. The runoff could not be diverted, because it was needed to hydrate the project site. In retrospect, this issue could have been avoided with coordinated planning and better project management.

One of the most contentious issues is controlled burning, owing primarily to liability concerns. If a project site requires intentional fire, then project managers can offer to burn an adjacent property concurrently on the basis that it would be beneficial to that property owner in terms of *fuel load* reduction or habitat improvement. However, an adjacent land owner could take legal action to prevent needed prescribed burns based on concerns of risk to personal property. In some cases, financial compensation can be provided to farmers or landowners for the right of restoration practitioners to burn around the perimeter of a restoration site in order to maintain a firebreak. One of us (JA) is engaged in a project in eastern Madagascar where this approach is being tested.

## Ecological Functionality

By ecological functionality we refer to the constellation of ecological processes that occur when living organisms interact with each other and their abiotic environment. Restoration practitioners ensure that the proper organisms occur in an ecosystem and that the abiotic environment supports them, so that they can grow, reproduce, enter into trophic interactions, and participate in numerous other ecological processes. Practitioners don't make organisms grow, reproduce, consume each other, or participate in those many ecological processes. Only the living organisms themselves can do this "work." This situation is analogous to turning on a light. We don't push electricity directly through the wiring and into the light bulb. Instead, we flip a switch to complete an electrical circuit. We also flip an ecological switch, so to speak, when we correct or adjust biophysical attributes, so that ecological processes can occur and reinstate functionality. For that reason, we recognize ecological functionality as an indirect attribute in table 5.1. Ecological processes in a restored ecosystem occur the same way that they did in that ecosystem before it suffered impairment. These processes occur at similar levels of performance, with allowances for the ecological stage of maturity and for any changes in environmental conditions.

Indications of functionality can be monitored in terms of the growth and reproduction of organisms, just as lumens can be measured as the output of a light bulb. All of these are indirect measures of functionality. Energy—the basis of all functioning—does not lend itself to direct measurement. If we could overcome the technical difficulties of setting up laboratory equipment outdoors, we might be able to conduct physiological investigations that would give added insight into the processes that eventually lead to growth. Such an exercise would be difficult, expensive, and beyond the technical capacity and budgets of most practitioners. Further, it would probably only yield limited information. Our only option is to search for indicators of functionality.

Indirect evidence may include measurements of plant size, vegetation cover, seed production, and vegetative reproduction. Soil organic matter can be monitored to detect increases, which indicate primary production and ensuing biological activity by detritivores. Plant vigor is inferred from an absence of stress indicators. Stress to organisms may appear in leaves that are discolored (chlorotic or infused with anthocyanin) or necrotic (with dead patches). Trees may defoliate early or exhibit dieback among the branches. In animals, abnormal behavior can signal stress. Lesions and tumors on fish are another indication of stress. The detection of such symptoms at a restoration project site is a signal that practitioner should search for causes and apply remedies.

Measurements indicative of functionality can be made rapidly by practitioners with modest training and outlays for equipment. A return to normal conditions in the physical environment generally indicates desirable ecosystem functioning. For example, increasing water clarity in streams may indicate that emergent vegetation is filtering suspended particulates from surface runoff. Increasing trends in diurnal measurements of dissolved oxygen can indicate recovery from eutrophication in an aquatic system. Temperature, pH, and electrical conductivity are easily measured with inexpensive equipment and reveal trends related to functionality. Water table elevation can be measured in piezometer wells. Stream discharge can be estimated from water depth data as recorded on a staff gauge or sometimes with recording instruments at weirs, as described in VFT 7 for temperate evergreen forest restoration in Chile. All of these are indirect measurements of ecosystem interactions.

## Historic Continuity

The reestablishment of historic ecological continuity means that biophysical conditions that were affected by impairment have been corrected and ecological processes have resumed. We have already described historic continuity in chapter 4 and emphasized that recovery from impairment does not necessarily mean that the predisturbance condition is regained. Instead, ecological restoration reconnects an ecosystem to its ecological trajectory and reestablishes the historic continuity that was temporarily interrupted. To employ a medical analogy, an impaired ecosystem is like a bedridden patient temporarily unable to get on with normal life. The biophysical conditions in need of correction are those associated with the first four ecological attributes listed in table 5.1. The reconnection will be smooth and seamless if environmental conditions have remained stable since the time of impairment and if corrections to biophysical properties can be made in accordance with the reference model. This degree of recovery is rarely attained fully, just as a medical patient may carry physical scars as reminders of injuries or relicts of surgical incisions.

Sometimes significant environmental change has occurred since impairment took place. Accordingly, the reference model, informed by current or anticipated changes such as arising from global warming, will specify biodiversity in the restored ecosystem, which contrasts substantially from that of the preimpairment state. This anomaly in biodiversity does not represent the termination of an ecological trajectory and the beginning of a new (or novel) ecosystem (see chap. 12). Instead, if impairment never occurred, and if the intact ecosystem were exposed to the same environmental change that actually occurred at the time when the ecosystem was impaired, then the intact ecosystem would evolve in response to the new environmental conditions. Perhaps the details of biodiversity would differ between an evolving intact ecosystem and the recovery of its impaired counterpart, but the direction of change would be similar enough to be interpreted as the continuation of the historic ecological trajectory.

## Ecological Complexity

Complexity and community structure are closely related, because structure without functionality is meaningless. This relationship gives rise to the concept of niche, which can be defined as structure in relationship to process. Technically, a niche is defined as the multidimensional space within an ecosystem that is occupied, and the ecosystem resources utilized, by populations of a particular species in a manner that allows individuals to participate in ecological processes. The concept of habitat is closely allied to structure and complexity. A habitat is the space or locality, and the ecosystem resources on which a species population relies. For an animal population, habitat is the space needed to search for food and water, seek cover from predation and environmental stress, participate in rituals of behavior, lay eggs, or rear young. For a plant, it is the locality in which a species or population naturally grows or lives—a peatland, a rocky cliff, a forest canopy, or the like. Complexity is the differentiation of that space or locality. Complexity makes it possible for ecological processes to accelerate or occur more efficiently, or to allow additional processes to occur. Niche is nuanced in a manner that emphasizes process, whereas *habitat* more nearly suggests structure. Habitat is loosely used by many authors as a synonym for ecosystem and for all the organisms that live there, but we reserve its use in a narrower and more meaningful sense that pertains to a single *taxon* or functional group. Ambiguity also arises when the term is applied to the natural geographical area of a taxon and to a locality where a specimen was collected.

In earlier seral stages at a restoration project site, structure is usually simple and easily understood. Restoration practitioners commonly initiate structural development during the course of making biophysical corrections. The dibble used in outplanting tree seedlings affects soil structure, and aerial portions of these same seedlings, though small, add an immediate increment of aboveground community structure. Each seedling contributes a unit of structure, and at the end of the day the structure of the incipient forest equals the sum of its parts—the seedlings. After a few years of growth, those seedlings have become young trees that contribute to primary production, produce humus, serve as nesting sites for birds, provide shade for fledglings in those nests, and contribute to microclimate. By now, the sum of the forest structure is much greater than the parts. The arboreal structure generates emergent properties that were not present earlier. Structure that produces emergent properties is what is called ecological complexity. Complexity usually, but not necessarily, develops after implementation tasks have been completed at a restoration project site.

Complexity ultimately depends on a comprehensive complement of species, ensuring sufficient materials that can develop into complex structures. In this regard, an experienced restoration practitioner searches for opportunities to facilitate complexity in terms of niche differentiation. For example, introducing logs onto restoration sites diversifies habitat and allows those species adapted to living in or next to decomposing dead wood to occupy niches that would not otherwise be available to them, as shown in figure 5.3 and VFT 6. This has become a standard technique in stream restoration, whereby the introduction of large woody debris stabilizes eroding channels, differentiates stream segments into pools and riffles, and provides ligneous habitat for a multitude of animals.

Another way that practitioners can facilitate complexity is to plant vegetation densely on stream banks. Roots of these plants bind the soil and keep it from collapsing when stream banks are undercut by water currents (fig. 5.5). The undercut banks add habitat and complexity by providing cover from predators for amphibians and reptiles. We caution, though, that the facilitation of niche differentiation requires forethought. For example, we know of an instance where restoration practitioners installed a brush pile to attract small mammals to a project site. The exercise was so successful that the abundant mammals stripped and ate all of the bark from newly planted tree seedlings. Forethought can avoid nasty surprises, for example, installation of the brush pile could have been delayed a year until trees had grown larger.

**FIGURE 5.5.** Incipient undercut bank from stream flow at Dogleg Branch restoration, Florida, where the roots of restored vegetation hold the soil together and prevent its collapse.

## Self-Organization

In chapter 3, we described several properties of intact ecosystems that are diminished on account of impairment. These included the capacity to conserve moisture and mineral nutrients, the ability to maintain a favorable microclimate, the prevention of trophic cascades, and various beneficial soil properties. The latter pertain to soil stabilization, aeration, vertical mixing, and accretion of organic matter, moisture retention, and trophic interactions among others. All of these are properties of self-organization that emerge as a consequence of the relationship between ecological processes and structure.

Self-organization is essentially the consequence of tightly operational feedback loops that increase ecological efficiency and stability within an ecosystem. A feedback loop develops when the results from one process reinforce or otherwise influence the results of a subsequent process. For example, oscillations in species populations between a predator and its prey represent the effects of feedback.

Increase in predators causes decrease in prey to the point that predators decrease from insufficient food, which allows prey to increase. In a complex ecosystem, where many kinds of feedback loops operate simultaneously, multiple feedback loops can affect a species population in different ways. Species populations are thereby tightly regulated. Such an ecosystem is likely to favor longevity in populations of K-strategists during periods of environmental stability.

Hutchinson (1959) suggested long ago that a community increases in stability as the number of links in the food web increase. The number of links is a function of species diversity: the more species, the more linkages, and the greater the stability. This principle reinforces our emphasis on the importance of ensuring that restoration provides a comprehensive species composition. Although it might seem obvious that a suite of functionally diverse organisms is needed to sustain ecosystem functioning, it is only in the last two decades that the relationship between biodiversity and ecosystem functioning has received intense scientific investigation (Reiss et al. 2009). The implications for ecological restoration have only begun to be explored.

## Resilience

Resilience, or the capacity of an ecosystem to self-recover from stress or disturbance, is a property that emerges from ecosystem functionality, complexity, and self-organization. Restoration practitioners can do little to improve resilience, other than to develop an effective project design and implement it expertly to remove inhibitors and apply triggers for recovery. Once prescribed manipulations have been applied, practitioners can only await ecosystem response to stress or disturbance to assess how much resilience a restored ecosystem expresses.

Occasional observations have appeared in the literature, which hint at the effectiveness of emergent attributes at project sites. We have observed resilience at project sites on mined and reclaimed land in the central Florida phosphate mining district. Hall Branch restoration is one such project where a headwater wetland forest was successfully restored (Clewell 1999). After restoration project work was completed, mining began on an adjacent property, which caused the water table to become substantially depressed beneath restored Hall Branch for two years. Unusually severe drought occurred concurrently. Soils dried out, became fissured, and organic matter was lost from oxidation. By the time that normal wetland hydrology returned, only a few trees had succumbed and the forest remained intact, although coverage was reduced for populations of some herbaceous species and several alien species colonized the site. Reconnaissance suggested that the young restored forested wetland ecosystem had developed a degree of resilience to inadequate hydration.

## Self-Sustainability

The ability of an ecosystem to persist and sustain itself indefinitely is the ultimate ideal of ecological restoration. Self-sustainability is another ecological attribute that emerges on account of ecological complexity, self-organization, and resilience. Restoration practitioners cannot restore self-sustainability, but they can design and execute projects with self-sustainability in mind. This is analogous to a child participating in sports and dreaming of receiving Olympic gold medals. That is a powerful dream, and so is determination to restore an ecosystem to the point of self-sustainability.

Self-sustainability does not imply perpetual autonomy. Instead, it means that the restored ecosystem will reach a level of sustainability that is comparable to an intact ecosystem of the same kind that occupies a similar landscape setting. If that ecosystem is semicultural, it will require periodic management, either from traditional land usage or more technical ecosystem management. Ecosystem management may also be needed to counteract anthropogenic impacts or to maintain the ecosystem in seral stages of development as desired by stakeholders. Self-sustainability does not infer stasis. Instead, a self-sustainable ecosystem is dynamic and has the capacity to evolve in response to changing environmental conditions.

## Biosphere Support

In addition to its inherent value and importance within a landscape and an ecoregion, all restored ecosystems contribute to biosphere support. We use biosphere support to designate those attributes of local ecosystems that have particular consequence to maintaining or improving environmental quality of the entire biosphere. Increases in photosynthetic activity that accrue from restoration generate atmospheric oxygen and absorb carbon dioxide. Organic carbon is sequestered in plant biomass and detritus, which reduces the potential for greenhouse gases reaching the atmosphere where they may cause climate change. The recovery of ecological complexity incrementally adds to biosphere complexity and thereby increases its resistance to anthropogenic climate change. Restoration increases an ecosystem's capacity for thermal reflectance thereby increasing the cooling efficiency of the planet (Clewell and Aronson 2006).

While we might be tempted to underestimate the importance of restored arid lands (relative to lush tropical vegetation) in an assessment of biosphere support, this would be a mistake. Successful ecological restoration augments plant biodiversity in arid regions (Bainbridge 2007), primary productivity, and overall ecosystem functionality and resilience. In a very recent, detailed empirical study that was carried out in arid areas of the world, Maestre and coworkers (2012) showed that the preservation of plant biodiversity is crucial to buffer negative effects of climate change and desertification in arid areas—which represent 41 percent of the land surface of the Earth. Evidence of this sort emphasizes the value of ecological restoration to biosphere support.

## Goals and Project Standards

Ecological attributes of restored ecosystems listed in table 5.1, and recapitulated in figure 5.1, could serve as goals for all ecological restoration projects everywhere. No other ecological goals are needed, although more can be added, such as the provision of specific ecosystem services that benefit people, or the recovery of habitat for a particular species. The ecological attributes in table 5.1 could also be adopted as standards of practice for the profession of ecological restoration. The first four attributes can be satisfied by direct interventions. The other seven attributes are indirect, and could be understood in terms of the intent of ecological restoration. Biophysical interventions would be planned and conducted in a manner that would be expected to lead to their eventual realization. Monitoring that suggested otherwise would trigger midcourse corrections or adaptive management.

Many and perhaps most projects will be unable to achieve all attributes satisfactorily on account of circumstances beyond the control of restorationists. Circumstances may include, among others, inadequate funding, unrealistic expectations, extreme weather conditions, regulatory constraints, insufficient time to provide needed aftercare following implementation, and impacts in the surrounding landscape over which practitioners lack control. An inadequate or ineptly prepared reference model, however, is not an unavoidable circumstance. A sufficient body of restoration experience exists for planners and managers to argue forcefully for the resources they need to ensure that a project and expectations for it are matched to the resources available and regulatory constraints. For example, a project may be stretched over longer timeframes rather than linking it to budget cycles. In this regard, Stromberg et al. (2007) described a broad continuum of effort that was needed to satisfy the recovery of increasing numbers of ecological attributes recognized by SER (2004) in the restoration of California grasslands. In spite of the desire to improve the quality of restoration projects, our attitude should be to accept with gratitude what was accomplished rather than to denigrate a project for unavoidable deficiencies. As mentioned, surprises will occur as well, even in the best-planned and best-financed projects. Restorationists must be prepared to cope with elements in ecosystems that remain beyond the range of our ability to predict consequences.

This chapter pertains to ecological attributes of restored ecosystems. Socioeconomic and cultural attributes are equally important, although they may not come to fruition unless ecological attributes are satisfied. Much is being written on this subject, and we shall not dwell upon it any more than we have in chapter 2. In the next chapter we explore ecosystems that have been shaped by culture and upon which people depend in terms of their socioeconomic values. In chapter 12, we discuss socioeconomic issues in the process of considering the way forward toward truly holistic restoration.

# Chapter 6

# Semicultural Landscapes and Ecosystems

Thus far, we have said little about the trajectories of ecosystems prior to their impairment, other than that they were once biophysically and functionally intact in the past. Some readers may assume that those intact states were essentially manifestations of Nature, devoid of human artifice. This chapter dispels that misconception. Humans have influenced the entire biosphere, and all ecosystems bear the marks of artifice. In some ecosystems, evidences of human influences are grossly apparent. In others, it takes a thorough knowledge of local natural history before the lingering imprint of human activity comes into focus. When we restore, we commonly incorporate ecological legacies derived from human pursuits that were acquired in the *creation* of a semicultural ecosystem.

The making of a semicultural ecosystem or landscape is not at all idyllic. The anthropological literature is filled with tales of environmental degradation and consequent collapse of tribes and societies (Diamond 2005; Mann 2011). Tribal people who were living off the land were concerned about survival and their next meal and not about the romantic fiction of living in peaceful harmony with the environment. When a human population size reaches the carrying capacity of the land, people are already contributing to its degradation. Migration was (and still is) a common solution, if famine, disease, and warfare had not already defused the problem. Although ecotopia, as described in chapter 2, remains an overarching goal of humanity, it will remain elusive until people learn to live within their environmental budgets. In the meantime, practitioners of ecological restoration will continue to clean up the ecological wreckage from human enterprise until we learn to respect the biosphere and each other.

Humans are wide-ranging creatures who roam across entire landscapes and sometimes much more widely. Landscapes heavily influenced by human activities can be called cultural landscapes. However, nearly all landscapes bear at least a few marks of cultural influence and, for that reason, cultural landscapes and natural landscapes should be seen as occurring along a continuum. A landscape that was entirely transformed to production systems or reallocated for other purposes would certainly merit being called cultural or socioeconomic. At the other end of the spectrum are landscapes that are organized by natural, nonhuman-mediated processes. Between these extremes are rural areas that display both natural and cultural influences and can conveniently be called semicultural landscapes. Component ecosystems that display both natural and cultural influences we designate as *semicultural ecosystems*. A cultural landscape may consist entirely of semicultural ecosystems. Alternatively, natural areas, production systems, and reallocated lands may also be imbedded in them. The term *socioecological* landscape can be used interchangeably with semicultural landscape, but it may be too technical for general acceptance.

People who reside in semicultural landscapes live on the land, managing it from within and serving in a stewardship capacity. Under those circumstances, there would seldom be any need to conduct ecological restoration or apply external ecosystem management. However, if semicultural landscapes are abandoned, they could be reallocated by other parties to more intensive land use, or the land could remain vacant and undergo unassisted transformation to an undesirable state. Under such conditions, ecological restoration could eventually become necessary.

Approaching an ecological restoration project of that sort raises questions that are peculiar to culturally influenced environments. What if the socioeconomic conditions have changed so radically that nobody wants to continue cultural practices of the past? What if semicultural ecosystems were transformed on account of disuse or were impaired by external influences that were not welcomed by local people? Should such ecosystems be restored to a more natural state without regard to past cultural practices? Who are the stakeholders to be consulted on such matters? Are they the traditional stewards or tenders of the land, or are they those who would disregard the past and implement a new regimen of land use? These questions are compounded by the realization that in the absence of sustainable traditional or local cultural practices, semicultural ecosystems do not necessarily revert to a former, all-natural state if they are not tended. Instead, many of them cross one or more ecological thresholds and switch to a less desirable state, perhaps irreversibly, unless intensive ecological restoration is undertaken. Under these circumstances, a substitute ecosystem may have to be selected as the target for restoration. Lamb and Gilmour (2003) have addressed these questions and described how abandoned semicultural ecosystems can be sustainably restored—or at least rehabilitated—in a manner that facilitates livelihoods for local people.

Practitioners who restore such lands should be familiar with traditional land usage, including fire, for restoring semicultural ecosystems. In particular, practitioners must distinguish between normal land use and disturbance from human exploitation and misuse. The distinction between them is not always easily determined. Semicultural ecosystems benefit from a mutual relationship between nature and culture, whereas other ecosystems in semicultural landscapes may have suffered ecological harm from ecologically unsound land use. Rogers-Martinez (1992) asserted convincingly that restoration of a semicultural landscape requires the concurrent restoration of culture and that the two are inseparable. This theme has been treated in depth by Janzen (1988, 1992, 1998, 2002), Bonnicksen (1988), House (1996), Higgs (1997), Farina (2000), Naveh (2000), Harris and Van Diggelen (2006), and Moreira et al. (2006).

## Culturalization of Landscapes

The transformation of natural ecosystems to semicultural states was instigated many millennia ago by people around the globe who killed megafauna (large animals) and intentionally ignited fires. Within a millennium or so after the migration of Clovis People across the Bering Strait into the Western Hemisphere nearly 13,000 years ago, thirty-one genera of megafauna had become extinct, including mammoths, mastodons, wooly rhinos, ground sloths, armadillo-like glypodonts, saber-toothed tigers, horses, camels, and others. Herbivores among these animals consumed enormous amounts of biomass. Their rapid extirpation must have sent ecological shock waves throughout the Western Hemisphere that changed patterns of biodiversity in ways we can scarcely imagine. McCann (1999) reviewed the explanations of how so few primitive people were able to cause such widespread extinction. The consequences of these extinctions on biomes and ecoregions helped initiate the process that culturalized the biosphere.

By far the most important cultural tool in creating semicultural landscapes was fire. We devote much attention to fire in this chapter, because of its importance in understanding semicultural landscapes and because restorationists need to be well versed in fire as a major tool of ecological restoration. Fire historian Stephen Pyne presented incontrovertible evidence for the preeminence of fire in the culturalization of landscapes. He wrote unequivocally that "Fire and humans have coevolved....Together they have repeatedly remade the earth." (Pyne 1995, 4). Fire was closely integrated in many stages of the evolution of early hominids (genus *Homo*) since they appeared in eastern Africa about 2.5 million years ago (Wrangham et al. 1999). Some scholars proposed that "the rise of *Homoerectus* from its more primitive ancestors was fueled by the ability to cook—that is, to use fire" (Pausas and Keeley 2009, 596).

McCann (1999), building on Pyne's work, asserted that until recently, humans would ignite any landscape that could be burned and used fire to exert control over landscapes to aid settlement processes. Easily burned ecosystems were settled first and nonflammable ones later. The consequence of burning has been to expand geographic areas of the world occupied by pyrogenic ecosystems and to stimulate the creation of new kinds of pyrogenic ecosystems. Although many fires are ignited by lightning (Komarek 1966) or, rarely, by other natural causes, human ignition is the likely source of most fires everywhere.

Ecological restoration practitioners use prescribed burning for two principal reasons. One is to reduce fuels, and the other is for ecological effect. Fuel reduction burns are intended to reduce the likelihood of conflagrations in areas where detritus has accumulated on account of fire reduction or intentional fire suppression. Curtailment of burning can eventually impair an ecosystem and cause it to switch to an alternative state. Fuel reduction fires combust accumulated detritus and with it the fuel needed to cause damaging wildfires.

Ecological fires are ignited to improve species composition and community structure. Unwanted fire-insensitive species are weakened or killed or their aerial biomass reduced. Ecological burning may begin with a fuel reduction fire to remove the bulk of accumulated detritus, so that more specifically directed fires can later produce a desired, ecological effect. Sometimes the fuel reduction fire is needed to expose a lower layer of wet detritus and allow it to dry out for future ignition. Another reason for an initial fuel reduction fire is to remove part of the fuel in weather conditions that will not allow all of the fuel to burn. If it all burned at once, the fire may be too intense and would kill desirable plants, or fire would kill trees whose roots had penetrated the lower layers of this detritus.

Fires remove competing vegetation, expose mineral soil, and provide a flush of soluble nutrients from ash, all of which promote plant establishment. Experimental studies in the Mediterranean region demonstrated that ash cover affects seedling emergence (Izhaki et al. 2000), at surprisingly fine resolution. Heat from fire sometimes stimulates seed germination (Martin et al. 1975). Smoke from fires initiates germination in a whole suite of Australian woodland species (Rokich et al. 2002) and for a number of Mediterranean plants as well (Crosti et al. 2006). An understanding of germination behavior in response to fire is an important consideration in prescribing fire for restoration.

Grasses and *sedges* are particularly flammable owing to the high density in which they grow and to the splint-like growth form of their leaves that allows atmospheric oxygen to flow around them. The flammability of pines and the leaves of other conifers is attributable to their high content of terpenes and other volatile substances. In contrast, many plants resist ignition, such as the leaves of many broadleaved hardwood trees that are lacking in volatile compounds and that tend to lay flat on the ground after they have fallen in a way that prevents much penetration by atmospheric oxygen. Knowledge of fuels and their flammability is important for planning restoration projects.

Pyrogenic ecosystems are commonly characterized by the frequencies of fire, which are expressed as the return interval or average number of years between successive fires. It would be more accurate to express the return interval in ecological terms, such as the time it takes for sufficient fuels to accumulate to carry a fire, which, in turn, reflects recent rainfall or site-specific conditions that influence growth rates of plants. Most pyrogenic ecosystems, such as longleaf pine savannas (figure 6.1), are maintained by surface fires, in which grasses and other fine fuels burn quickly, and lethal temperatures only reach one or a few meters above ground level. In frequently burned savannas, trees are generally adapted to survive such fires, and their wide spacing prevents crown fires from jumping from one tree to another. In other pyrogenic ecosystems, the return interval may extend for decades, and fires consume dense, tall shrubby undergrowth and ignite the crowns of trees, sometimes killing them. Examples are pocossins (evergreen shrub-bogs) and bay swamps in southeastern North America, jack pine (*Pinus banksiana*) forest in the Great Lakes Region, lodgepole pine (*Pinus contorta*) forests in Western North America, *Eucalyptus* forests in Australia, and various Mediterranean-type woodlands in southwestern Australia and the cape region of South Africa.

**FIGURE 6.1.** Longleaf pine savanna in central Florida, USA.

Nearly all grasslands and savannas are pyrogenic ecosystems. Many large mammals and birds with which people are familiar are residents of those ecosystems. Those animals depend on fire for their survival, in spite of occasional fictions to the contrary, such as Walt Disney's famous movie, *Bambi*. Restoration practitioners need to know whether or not an impaired ecosystem is pyrogenic, what the return interval is, and what kind of fire behavior is normal for it. This kind of information is incorporated into the restoration model and into recommendations for postrestoration management that practitioners may be required to prepare. The substitution of herbicidal application or mowing is commonly suggested as a substitute for prescribed fire in restoration projects. This expedient may avoid the hassles of obtaining burning permits or the costs and risks of liability posed by prescribed burning. However, mowing without subsequent burning is ecologically unacceptable, if mowing causes a litter layer to form which suppresses fire-adapted species. Herbicides, or at least those with longer residence times in the soil prior to degradation, may find their way into groundwater and contribute to environmental pollution. By contrast, burning removes litter, releases mineral nutrients quickly in soluble ash that is absorbed by shallow root-mats of the fire-adapted grasses and *forbs* and is expressed shortly after the next rainfall event by a flush of verdant new growth.

## Examples of Semicultural Landscapes

We now examine examples of semicultural landscapes from around the world. Semicultural landscapes are widespread and occur wherever people reside. Most ecosystems and inhabitable landscapes are semicultural in those regions of the world that have been densely and continuously populated by people for several millennia, whether or not traditional land and resource uses are still practiced. An assessment of semicultural ecosystems in California, USA, by ethnobotanist Kat Anderson (2005) reveals a much greater prevalence of *cultural ecosystems* and landscapes than had been previously recognized in the Western Hemisphere. California is large, dramatically varied topographically and climatically, and supports a wide range of ecosystems. Anderson's work shows that most of these ecosystems were culturally altered and that the only entirely wild ecosystems in California were subalpine forest, deserts, salt marsh, beach and dune ecosystems, alkali flats, and serpentine outcroppings. Anderson (2005, 8) claims that, "much of what we consider wilderness today was in fact shaped by Indian burning, harvesting, tilling, pruning, sowing, and tending." She argues eloquently (158) that "much of the landscape in California that so impressed early writers, photographers, and landscape painters was in fact a cultural landscape, not the wilderness they imagined. The wildflower displays they depicted were edible plant gardens." For example, open and grassy oak savannas of California's valleys and foothills were entirely shaped by intentional burning and by the tending of valley oak (*Quercus lobata*) for the food value of their acorns and a variety of other uses. Supporting evidence for this contention was provided by McCarthy (1993), who showed that in the absence of fire, coniferous trees encroached into these savannas at the expense of oaks. Extensive coastal prairies were largely coniferous forests in the past that were transformed by Indian burning and subsequently fire-maintained by them as both a source of food from wild plants and as a hunting ground for native animals (Anderson 2005).

The prevalence of cultural landscapes in California should not come as a surprise. Day (1953) had documented similar ecological influences by American Indians in the northeastern United States half a century earlier. University of Wisconsin geographer, William Denevan (1992) presented compelling arguments to dispel what he called the "pristine myth" of virgin landscape in the Americas at the time of the first voyage by Columbus back in 1492. That myth, which he claimed had been woven by nineteenth-century romanticists and "primitive" writers such as Cooper, Longfellow, and Thoreau came to be accepted as undisputed fact until very recently. Denevan (1992) estimated a substantial indigenous human population in the Western Hemisphere of 40–100 million people in 1492. He claimed that this population was then reduced by European-borne diseases by 89 percent a mere 158 years later, by 1650. By that time, however, through secondary biological succession, the indigenous biota had sufficiently recovered from sustained cultural land use so as to suggest wilderness to European colonists.

Denevan (1992) emphasized that American "Indian impact was neither benign nor localized and ephemeral, nor were resources always used in a sound ecological way." Over several millennia, Native American Indians had developed semicultural landscapes by the time of Columbus's arrival. Agricultural clearing and burning had converted much forest into fallow growth and into semipermanent meadows, glades, savannas, and prairies.

Some of the earliest and strongest evidence of intentional, prehistoric burning comes from Australia. Among many other scholars, Bowman (1998) asserted that humans colonized that island-continent 40,000 years ago and perhaps earlier. Palynological data from lake sediments showed abrupt increases in the frequencies of fine charcoal particles and of pollen from *Eucalyptus* (figure 6.2) and other fire-tolerant trees that were more easily explained on the basis of aboriginal-set fires than by climate change and ignition. Bowman wrote that there was little doubt that aboriginal burning was central to the maintenance of many or most Australian landscapes long before the time of European contact. This burning caused substantial changes in the geographic range and demographic structure of much vegetation, not to mention the extinction of many kinds of plants and animals.

**FIGURE 6.2.** *Eucalyptus* forest near Sydney, Australia, with the undergrowth dominated by woody species.

Fires favored the expansion of grass and shrub lands, and they allowed *Eucalyptus* to colonize rain forest. Flannery (1994) argued that Aborigines used what is called "firestick farming" in Australia, which altered Australian ecosystems. This practice set off a coevolutionary process that Flannery and others claim has contributed to the nutrient-poor soils and strongly fire-adapted ecosystems in all but the wettest northeast corner of the continent. Flannery (2001) contended that firestick farming was also practiced on the North American continent. This cultural history raises perplexing questions for restorationists regarding the wisdom of restoring historical pyrogenic ecosystems in a time when needs and choices have changed for most Australians, including those of the now-sedentary Aborigines who live in a very different manner than did their ancestors a few generations ago.

Much of the Amazon Basin consisted of a semicultural landscape at the time of European contact; the only question is how much was relatively untouched by human influence. Clement (2006) estimated a population of four to five million people for Amazonia at that time. In 1542, Spanish explorer Gaspar de Carvajal described the banks of a 180-mile stretch of the Amazon River as being densely inhabited with essentially contiguous villages (in Mann 2005). Balée (2000 and references therein) described forests growing on enormous mounds of potsherds in the Amazon. Clement (1999) identified 138 plant species that were cultivated or managed by tribal peoples. Many of these species were trees bearing fruits or other edible parts, which were planted or seeded throughout the Amazon, thereby converting it into a vast agroforest. Ongoing work on ecology and archaeology in Belize (Ross 2011) and on coastal savannas of French Guiana (McKey et al. 2010) strongly support the theory that those landscapes were "co-constructed by man and nature."

Lomas, or fog oases, (Rundel et al. 1991) constitute regional hot spots of biodiversity that extend hundreds of kilometers on the lower Pacific slopes of the Andean Cordillera (Puig et al. 2002). Surrounded by one of the driest regions on Earth—the coastal desert region of northern Chile and southern Peru—the lomas are sustained by fog that is intercepted by a tree canopy dominated in part by tara (*Caesalpinia spinosa*). This forest has suffered 90 percent reduction within the past century. A multidisciplinary study is under way to design a restoration program (Balaguer et al. 2011). Genetic and ecophysiological evidence suggests that tara was intentionally introduced to lomas before Europeans arrived because of its usefulness to indigenous peoples as a source of tannin and natural dyes. Therefore, Balaguer et al. (2011) recommend that restoration strategies for lomas should include agroforestry practices to emulate the way they were managed in the past and in reflection of the way they evolved as semicultural ecosystems.

Semicultural landscapes are of course very common in Europe, such as moors that were deciduous forests in Neolithic times, now reduced to a low dense scrub with greater species diversity than might be anticipated in a cold-temperate climate. Chalk meadows, described in chapter 4, constitute another semicultural landscape. One dramatic example is centered on cork oak (*Quercus suber*), a long-lived, multipurpose tree that is distributed throughout much of the western Mediterranean region and is highly prized for its bark from which bottle stoppers and myriad other products are fabricated. It seems likely that people have transported its acorns since prehistoric times, expanding its range throughout southwestern Europe and northwestern Africa, where it now provides an ecological framework for open woodlands known, among other names, as *dehesa* or *montado*. These are semicultural agro-silvopastoral woodlands that are maintained by a strong human input, including fire and the intentional inclusion of other species. For example, pines are planted in cork oak woodlands in Portugal (Pausas et al. 2009). In localities where woodlands were abandoned and fire management was curtailed, the cork oak woodlands have grown into highly flammable forests and pose a danger of conflagration. Some say the danger could be averted by converting these sites to a less flammable forest type, but others see the value in restoring them for their heritage value, wildlife value (they provide habitat for endangered birds and lynx), and agritourism. Cork oak woodlands suggest that public policy decisions may be needed to determine the disposition of abandoned semicultural landscapes.

## Selecting Semicultural Targets for Restoration

More than one semicultural ecosystem can evolve from its "wild" state in response to different cultural practices or different intensities of practice. In other words, multiequilibrium theory (chap. 4) may apply to the selection of a target ecosystem for ecological restoration. For example, the tallgrass prairie of the North American Midwest is maintained by frequent fires. At the time of European contact, that ecosystem may have owed its widespread occurrence to annual burning by American Indians. In addition, some areas were prairie-like but were interspersed with occasional large individuals of burr oak (*Quercus macrocarpa*) and several herbaceous species that occurred nowhere else (Packard 1988, 1993). These burr oak savannas are recognizable as an alternative state that was presumably the product of a fire regime more irregular than it was in tallgrass prairies. In the greater Chicago region, which extends into Wisconsin and other nearby states, curtailment of fire has allowed deciduous forest to replace tallgrass prairie and burr oak savanna. We may never know which of these three kinds of ecosystems represented the wild state from which the other two developed as alternative states. Post-Pleistocene climates have fluctuated subsequently and obfuscated whatever evidence may be brought to bear on this question. Any alternative state is valid as a restoration target, and stakeholder consensus would be the ideal method for its selection.

In another example from eastern North America, forest dominated by American beech (*Fagus grandifolia*) and sugar maple (*Acer saccharum*) represents the state of ecological equilibrium or climax on *mesic* soils in northern Ohio. American Indians were once prevalent in this region and routinely ignited surface fires that removed undergrowth and killed smaller trees. This practice favored forest that was codominated by white oak (*Quercus alba*) and hickory (*Carya* spp.). Karl Smith conducted prescribed fire in beech-maple forest to restore this oak and hickory forest as a semicultural ecosystem. Burning stimulated a profusion of spring wildflowers, whereas these same herbaceous plants had previously persisted as dormant rootstocks beneath the formerly dense cover of beech and maple (Smith 1994). Similarly, small fires have been ignited in patches to diversify the composition and structure within old-growth "climax" forest in southern Illinois by Stritch (1990), and in adjacent Missouri by McCarty (1998).

In the Mediterranean region, alternative ecosystem states reflect variation in the cultural fire regime, the intensity and frequency of wood extraction through coppicing or pollarding of trees, and the intensity of grazing by domestic livestock. Plant species composition is similar in every alternative state, but the relative abundance of the species and thus the community structure varies markedly among them. The least cultural activity allows oak forest to develop, which may represent the wild state. However, the duration and intensity of human occupancy in this region cast doubt on any purely natural state ever being identified with certainty. Fire and fire-related activities can transform oak forest either to pine woodland or to an oak-dominated shrubland known from one region to another as *garrigue*, maquis, or matorral (for synthesis of a large literature, see Blondel et al. 2010). Garrigue is further transformed in structure by coppicing, which is a local cultural practice. Frequent burning, often coupled with livestock grazing and fuelwood collection, transforms garrigue to shrubland dominated by *Ulex*, or into sward, which is a type of grassland. Restoration can be modeled after any of these alternative states.

The selection of the preferred alternative state in restoration planning is critical, because stakeholders and members of the broader community will have to live with the results, and sometimes appearances can be deceiving. We recommend the following guidelines for the selection of which alternative state to choose as a reference system or short-term target for restoration, if several exist. First, be practical! Preference should be given to those states for which any required ecosystem management following the completion of restoration could effectively be provided without taking extraordinary measures. If, for example, an alternative state requires frequent prescribed fire and burning permits will not be issued at any time in the near future, then that alternative state should not receive consideration as a restoration target. Second, if the restored ecosystem will be utilized in some way, such as for recreation or aesthetic improvement, only alternative states should be considered that are conducive to that purpose. Third, if long-term sustainability is a goal of restoration, a relatively stable and persistent alternative state should be selected.

We may ask, what are the ecological consequences of ecosystem transformation by traditional cultural practices to alternative states—or metastable states as they are sometimes called? The most frequent consequence is that forests or other closed communities are converted to more open states, such as a woodland, savanna, or grassland. Our description of opening up forests in Ohio is an example. Ecologists and natural area managers have all too often ignored or discounted cultural practices as equivalent to climatic factors and sometimes other entirely natural ecological drivers. The traditional assumption has been that these more open states represent earlier stages in biotic succession that were caused by human-mediated disturbance. Preferences of local communities have been ignored—at least in North America, India, and Australia—in deciding between protected areas on the one hand, and stewardship by people living close to the land on the other. It now appears that local people need to appreciate and benefit directly from the various services a park or other protected areas can provide (Terborgh et al. 2002; Figueroa and Aronson 2006).

R. Marius van der Vyver, Richard M. Cowling, Anthony J. Mills,
Ayanda M. Sigwela, Shirley Cowling, and Christo Marais

The subtropical thicket biome is characterized by a tangle of spiny, often suc-culent trees and shrubs, with a canopy height typically reaching 2–5 meters (photo 1). Based on its rich array of endemic plants—notably succulents and bulbs—subtropical thicket forms the southwestern part of the Maputaland-Pondoland-Albany biodiversity hotspot (Steenkamp et al. 2005). Subtropical thicket was only recently recognized as a distinct biome; it is phylogeneti-cally and functionally related to tropical forest and predates the evolution of desert, savanna, and grassland biomes (Cowling et al. 2005).

Species composition and structure in subtropical thicket vary in response to an environmental gradient related primarily to annual rainfall, extend-ing from the arid west to the mesic east (Vlok et al. 2003). The more arid areas (precipitation < 450 millimeters year-1) are mostly dominated by a

**PHOTO 1.** Undisturbed subtropical thicket.

**PHOTO 2.** Densely planted Spekboom.

**PHOTO 3.** Aerial photo of subtropical thicket that is undisturbed on the left side and severely degraded from grazing on the right side.

succulent canopy tree, Portulacaria afra Jacq, known locally as Spekboom or Igwanishe (photo 2). Spekboom possesses drought-resistant characteristics such as CAM-photosynthesis, which allows it to survive and flourish even in times of severe water limitation. In addition, it is able to switch to C3 photosynthesis when rains come (Guralnick et al. 1984a, 1984b). This versatility may explain its high productivity relative to other succulents. Spekboom readily reproduces vegetatively from broken branches, and is highly palatable to wild ungulates, elephants and black rhino, and also for Angora and other domestic goats.

More than a century of overintensive goat browsing and unsustainable management practices have eliminated Spekboom populations in which other long-lived woody canopy trees and shrubs once flourished (photo 3). Of the 1.4 million hectares of thicket dominated by Spekboom, only 200,000 hectares remain (Lloyd et al. 2002). In its place is a barren, desertified, savanna-like landscape with some remnant long-lived woody canopy species, exposed soils, ephemeral herbs and grasses (called "opslag"), and karroid shrubs uncharacteristic of the former intact thicket landscape (Hoffman and Cowling 1990). Fenceline contrasts provide clear evidence of the impacts of overstocking on Spekboom-dominated thicket.

Consistent with the alternative stable states model of ecosystem dynamics (Milton and Hoffman 1994), these pseudo-savannas are on a trajectory toward a new stable state. In other words, they are fast becoming a treeless karoo (semidesert) landscape as there is neither recruitment of any original thicket canopy species, nor of Spekboom. The remaining long-lived thicket trees and shrubs show diminished reproductive output and greater mortality compared to those flourishing in intact thickets (Sigwela et al. 2009).

Where Spekboom prevails, it is the major contributor to the thick litter layer beneath the canopy (Lechmer-Oertel et al. 2008), and its own dense canopy with the characteristic basal "skirt" provides a suite of microenvironmental conditions, such as cooler temperatures, improved infiltration rates, retained soil moisture, nutrients, and carbon (Mills and Fey 2004; Lechmere-Oertel et al. 2005; Mills and Cowling 2010). These are conditions necessary for the recruitment of woody canopy dominant species, as well as other subcanopy components characteristic of intact Spekboom-dominated thickets (Sigwela et al. 2009).

Sites where local landowners planted Spekboom truncheons (stem cuttings) to control soil erosion have provided opportunities for us to investigate the properties of decades-old restored sites, versus degraded and intact sites

**Photo 4.** Planting Spekboom truncheons at a restoration project site.

(photo 4). Spekboom establishes readily in water-limited degraded sites and over time, ameliorates their harsh conditions toward that of the intact state. Once it is reintroduced, Spekboom transcends an abiotic threshold and stimulates recovery of an intact and functional subtropical thicket ecosystem. Different-aged plantings of Spekboom truncheons revealed that thicket plant biodiversity regenerates spontaneously after a period of about thirty-five years postrestoration (van der Vyver et al., forthcoming). This finding allayed concerns that simply planting Spekboom truncheons would lead to species-poor Spekboom monocultures. It also reinforced the feasibility of this simple method given the high costs and impracticality of restoring species in addition to Spekboom (van der Vyver et al. 2012).

Local communities surrounding the dry subtropical thicket region are among the most impoverished in South Africa. Would ecological restoration of Spekboom thicket provide benefits to improve the lives of local residents? It was already known that subtropical thicket possesses high carrying capacity for livestock and game. It improves soil stability, and it increases soil water retention and flood attenuation. In addition, we discovered the ability of Spekboom to sequester carbon at a rate comparable to many mesic forest ecosystems (Mills et al. 2009), which could allow us to tap into the global carbon market in order to finance restoration efforts. We compared soil carbon under intact vegetation to restored sites and measured substantial amounts of sequestered carbon beneath Spekboom plantings (Mills and Cowling 2006).

Until recently, the region lacked social capital and external assistance of the sort needed to establish restoration programs and community upliftment. The South African government had been approached to fund landscape-scale restoration of subtropical thicket. In 2004, the Spekboom restoration project was adopted by the Natural Resources Management Programmes (NRMP), a multidepartmental initiative administered by the Department of Environmental Affairs as part of its contribution to the Expanded Public Works Programme (EPWP) aimed at alleviating poverty by providing additional work opportunities coupled with skills training. Thus was born the Subtropical Thicket Research Programme (STRP), directed by scientists and economists, and managed by the implementation agency, the Gamtoos Irrigation Board (GIB). The project is run on a tender-based system where a contractor and a team of workers are paid for planting a designated area according to standard specifications.

The goal of the STRP is to create a sustainable rural economy where thicket restoration facilitates new income streams via carbon credits and payment for other ecosystem services (Marais et al. 2009; Mills et al. 2009). The planting contracts provide training programmes for contractors that include skills for managing their own businesses involving tendering, bookkeeping and financial management, as well as technical skills related to restoration. Training for the worker teams includes the necessary technical skills and life development skills, such as personal finance, HIV training, and primary health care. In this way, the ecological restoration of the landscape is implemented through the work of the poorest of the poor, namely unemployed local people who are able to benefit directly from restoration (Mills et al. 2009). Ecological restoration of subtropical thicket, as guided by the STRP, promises to make major regional recoveries of biodiversity and people's well-being.

Pedro H. S. Brancalion

In the state of São Paulo, southeastern Brazil, large areas of very high-diversity Atlantic Forest have been converted to agricultural production. In the last two centuries, forest cover was reduced from 85 percent to around 10 percent, mainly as a result of deforestation for coffee plantations. In the last decades, sugarcane fields have assumed the main agricultural position and cover about five million hectares of land. Sugarcane production depends on highly mechanized cultivation of the soil and the use of fire before manual harvesting. These practices facilitate erosion and the siltation of water courses, particularly where sugarcane is grown on riparian buffers. Silt accumulates behind dams on rivers, and reservoirs upstream of these dams supply little or no drinking water in dry seasons.

An unfortunate instance of siltation occurred in the reservoir that supplies the municipality of Iracemápolis with drinking water. Silt, which eroded from sugarcane fields that directly adjoined the reservoir, filled it to the point that it did not contain water during a severe drought in 1985. For several weeks, trucks were hired at great cost to bring water from neighboring cities.

To prevent recurrence, the city council, along with universities and outreach agencies, prepared a plan to elevate the height of the dam, dredge the reservoir, and expand its water storage capacity. A new shoreline and riparian zone would be formed on open agricultural land. The initial plan was to rehabilitate this zone by planting seedlings of tipuana (*Tipuana tipu*), an exotic tree. Tree planting would reduce erosion and set a definitive border beyond which sugarcane could not be planted. After further consideration, professors Hermogenes Leitão Filho of Campinas State University and Ricardo Ribeiro Rodrigues of the University of São Paulo proposed a visionary ecological restoration project to plant a forest consisting of 108 species of native trees. The proposal was accepted and was implemented between 1988 and 1990. Trees were grown in pots in a nursery and manually planted. Some of the more common tree species were *Piptadenia gonocantha*, *Centrolobium tomentosum*, and *Pterocarpus violaceus* (all legumes); *Cariniana legalis* (Lecythidiaceae); *Balfourodendron riedelianum* (Rutaceae); and *Handroanthus chysotrichus* (Bignoniaceae).

Today, this 50 hectare, high-diversity project site is one of the most successful restoration projects in southeastern Brazil. It has served as a model for

**PHOTO 1.** View of the restored forest from across the reservoir in 2012.

**PHOTO 2.** Interior of the restored forest in 2012.

restoration in human-dominated landscapes of the Brazilian Atlantic Forest region, the most threatened biome of the country. Undergrowth species, particularly shrubs and climbers, have colonized spontaneously. Epiphytes did not colonize as a result of dispersal limitations and a lack of suitable regeneration microsites. In order to accelerate the reestablishment of this important plant group, orchids, bromeliads, and *Rhipsalis* spp. (Cactaceae) were reintroduced by attaching young seedlings to the trunks of trees (photos 1, 2, and 3).

**PHOTO 3.** Woody vine that colonized spontaneously.

Amateur ornithologists have recorded more than sixty bird species in the restored forest. Native mammals observed in the forest include tayra (*Eira barbara*) and cougar (*Puma concolor*), both protected by CITES, appendix 2, and squirrel (*Guerlinguetus ingrami*), coati (*Nasua nasua*), possums (*Didelphis* spp.), capibaras (*Hydrochaeris hydrochaeris*), armadillo (*Dasypodidae* spp.), and South American river otter (*Lontra longicaudis*).

Although the population of this municipality has increased in recent years to 18,000 inhabitants, the drinking water supply has remained adequate. In addition to the role played by the project to contribute an ample supply of clean water and to increase and protect native biodiversity, this restored site has also been used as a "living" laboratory for environmental science classes and research projects. The site is visited by 755 students per year, including 350 local high school students, and by 305 undergraduate and 100 graduate students from the Luiz de Queiroz College of Agriculture, University of São Paulo. Approximately twenty scientific investigations have been conducted there. Additional benefits include aesthetic value in contrast to adjacent sugarcane fields, recreation, inspiration, ecotourism, and use for religious rites by residents of Iracemápolis. Approximately 3,000 people use the area for fishing, hiking, biking, and bird watching (photo 4).

**PHOTO 4.** Students attend field trip at the project site.

In an interview carried out with 300 people, 88 percent responded that they would like to take a guided tour to learn about the project, and 62 percent would pay for the tour (most of them, US$3). In addition, 87 percent believed that the restoration project improved the quality of the water they drink because of its reduced sediment and pesticide content. An impressive 94 percent of respondents wanted more forest restoration projects to be conducted in their community, and 63 percent would accept an increase (~ US$3 per month) in water costs to finance such projects. This amount would pay for approximately 30 hectares per year of forest restoration. However, 56 percent of the respondents specified that they would accept an increase in water cost on the condition that restoration projects would be locally administered. They would reject water rate increases for projects that were carried out by regional or national authorities.

Monitoring data from this project helped formulate public policies and specific legal instruments in support of restoring high-diversity tropical forests in Brazil, including Resolution 08 of the São Paulo State Environmental Secretariat, which strongly mandates the recovery of native biodiversity (Aronson et al. 2011). This project exemplifies how forest restoration benefits people and convinces stakeholders to approve societal investments in additional projects.

# PART III

# How We Restore

Part 3 addresses how ecological restoration projects are constituted—the nuts and bolts, if you will. Chapter 7 addresses reference sites and the preparation of *reference models* from which all planning flows. Here we note that the reference model is not necessarily static but can change with time, and can be conceived in a series, from the outset. Chapter 8 introduces a dilemma regarding the intensity of restoration work needed to ensure that ecological attributes are attained without compromising the naturalness of the ecosystems we try to restore. Then we begin to examine the sources of knowledge we tap into when we restore: is it an art, a craft, or a science? We talk about innovations, tools, and general approaches that may apply best to different types of ecosystems. In chapter 9, we identify all of the steps that are involved in any ecological restoration project. They begin with project conceptualization, and continue with the different phase of planning and implementation, and end with postimplementation tasks, evaluation, reporting, and celebration.

# Chapter 7

# Ecological References

Artists who paint landscapes don't work in cubicles in office buildings. They depend on natural scenery for inspiration and specific detail. So too with ecological restorationists who necessarily begin with natural references as the basis to conceive and formulate their projects. This chapter describes the various kinds of ecological references and how the information they contain can be synthesized into reference models for the preparation of restoration project plans. Later, we relate ecological references to ecosystem trajectories and explore the temporal aspects of ecological references.

## Reference Concept

An ecological restoration project begins with a representation from nature that guides all aspects of project planning and implementation. That representation is called the *ecological reference*, and it reveals ecosystem states and indicates what is known about underlying processes. Ideally, information revealed by an ecological reference includes species composition, community structure, physical conditions of the abiotic environment, exchanges of organisms and materials that occur with the surrounding landscape, and anthropogenic influences in semicultural ecosystems. Such biophysical information allows us to reinitiate or accelerate ecological processes that had been arrested or retarded.

The ecological reference can assume many forms and can be prepared from both primary and secondary sources of information. Primary sources are actual ecosystems, called *reference sites*. Written ecological descriptions of reference sites also qualify as primary sources, as long as they contain adequate information for preparing restoration plans. Secondary sources consist of any other information that contributes in some way to the description of an ecosystem prior to its impairment. More specifically, an ecological reference may consist of the following:

- an ecological description of the ecosystem to be restored before it was impaired
- remnants of that same ecosystem that survived impairment
- another intact ecosystem of the same type in the general vicinity
- a combination of the above elements or their ecological descriptions
- any of these choices with additions of secondary information or with modifications specified to accommodate changing or recently altered environmental conditions or constraints
- synthesis of secondary information in instances when reference sites or their ecological descriptions are unavailable

Preferably, the ecological reference will be assembled into a coherent document called the reference model that synthesizes all of the information that is needed to design and plan an ecological restoration project.

The reference for restoring xeric longleaf pine savanna (VFT 1) and subtropical thicket (VFT 4) consisted of intact remnants of original ecosystems. A thorough ecological description of Mediterranean steppe was available as the reference for VFT 2. The reference for shola grassland restoration (VFT 3) and Brazilian forest restoration (VFT 5) consisted of secondary information from regional floristic studies.

An ecological restoration project begins with a mental image or vision of how an impaired ecosystem would be expected to appear following restoration. In most instances, the vision reflects one or more reference sites that usually consist of intact ecosystems of the kind to be restored. The vision, whatever its inspiration in nature, is commonly called the restoration target and represents the intended long-term outcome of ecological restoration. With the vision in mind, the goals of an ecological restoration project can be formulated. These goals are succinct statements of intent, preferably prepared in writing, which identify the vision for the restored ecosystem and the socioeconomic, cultural, or personal values (chap. 2) that the restored ecosystem is intended to fulfill. Goals are broader in scope than the vision, because they consider dynamic as well as visual aspects of the ecosystem after its anticipated restoration, in terms of ecological processes and ecosystem services.

Ideally, the principal ecological goal of any restoration project is to reinitiate ecological processes in a manner that reestablishes historic continuity and that leads to the development of the ecological attributes of restored ecosystems (chap. 5). Other project-specific goals can be proposed by project sponsors, stakeholders, or anyone else who is involved in project conceptualization. These additional goals may concern desired ecosystem services and protection of imperiled species, among other values. The development of goals can be a life or death matter, as, for example, when a goal of restoration is intended to stabilize a mountain slope and prevent ruinous landslides from descending on people living below.

When no intact reference sites exist, particularly in those localities with long histories of intensive human land use, the vision for restoration and the reference model must rely on secondary sources of information. Dependence on secondary sources inevitably requires a measure of imagination and professional judgment, based on bioregional knowledge and experience, for its conception. However imperfect it may be, that vision and the reference model that describes it in technical terms serves as a valuable tool to guide restoration (Aronson et al. 1995; Swetnam et al. 1999; Egan and Howell 2001). Despite what some authors have suggested, a reference ecosystem or model is neither an instruction manual nor a straitjacket for restoration; instead it is a beacon and a pointer to the future.

The selection of the reference is rightfully the responsibility of stakeholders and members of the local community, who will necessarily live with—and provide stewardship for—the ecosystem that is restored. Once restored, that ecosystem should serve as a socioeconomic and cultural resource for peoples' benefit, which they will respect, protect, and manage, either directly or through institutions dedicated for that purpose. If the goal of restoration is not embraced by the local community, then the project should be reconceived or the project site reallocated for another purpose. Once the ecological reference has been selected, it needs to be described and conceptualized as a reference model in a manner that allows restoration plans to be developed, as will be described at length later.

The vision or target of an ecological restoration project is not necessarily attained at the time restoration project work has been completed. In many projects, restoration tasks are completed decades or centuries before the target state can be achieved. A restored coral reef might take millennia to recover to its target state, assuming environmental conditions remained sufficiently stable for ecological maturation to occur. The vision or target, as reflected in the reference model, may never be fully achieved. This should not be a cause for concern. The reference model is the starting point for ecological restoration and not the endpoint. Ecological restoration is openended. We cannot insist that the endpoint matches our preproject vision, or the product would not be natural, as explained in chapter 4 and illustrated with the analogy of a grounded ship. Once restored, we cannot be certain of the future course of the vessel or an ecosystem. A proper vision is essential, though, because it portrays the intent of restoration and provides anyone who is interested with a glimpse of the anticipated outcome.

Many completed restoration projects emulate their reference sites rather faithfully, and sometimes the waiting period is brief. Figure 7.1 gives an exceptional

**FIGURE 7.1.** Wet prairie restored in Florida. A sod cutter removed turf intact from a donor site for transfer to the project site. This photo was taken several weeks after transfer.

example of instant restoration, whereby an ecosystem—a wet prairie—was lifted from a site scheduled for mining as sod and transported directly to a restoration site on previously mined land. The physical environment was essentially identical to the donor site in terms of soils, hydrology, and topography. All that was needed was for plants in the transported sod to send roots into the mine soil, which happened promptly, as if a new lawn were being sodded in a residential development. After several weeks, the only evidences of restoration were faint lines where the strips of sod came together. In this case, the reference literally became the restoration project. The technique is not new. Munro (1991) has successfully transported wetland sod in the mid-Atlantic region of the United States.

In many projects undertaken in the past, no formal reference model was prepared, and an undocumented, often unarticulated target ecosystem informed restoration activities. This practice was satisfactory as long as impairment was not severe, if the project was limited in size, and if the practitioner was experienced in local natural history and served as the project planner. These conditions assumed a prolonged period of aftercare that followed project implementation, which allowed ample time and resources to nurture the recovering ecosystem and make any needed midcourse corrections. Ecological restoration is too important and too costly to continue this practice. We strongly advise the preparation of a reference model for every restoration project, however imperfect and incomplete that model may be.

The search for historical reference sites is made increasingly difficult on account of land usage and disturbance that obscure the characterization of original ecosystems. Brewer and Menzel (2009) developed a protocol using multivariate statistical ordination that can substitute when no suitable reference sites are available. The method relies on a species-by-habitat data matrix generated from biodiversity surveys that is intended to statistically assess similarity among communities. The preparation of reference models in the manner developed by Brewer and Menzel (2009) represents an opportunity for restoration ecologists to collaborate with practitioners and contribute significantly to the refinement and success of ecological restoration projects.

## Types of Reference Sites

White and Walker (1997) proposed a formal classification of reference sites consisting of four categories. They are (1) same place, same time; (2) different place, same time; (3) same place, different time; and (4) different place, different time. In the first case (same place, same time), the ecosystem to be restored contains sufficient evidence of its prior, intact condition to serve as its own reference, and it is called an auto-reference. In the second case (different place, same time), the primary reference site is called a refuge, indicating that a portion of the ecosystem remains intact and can serve as a reference for other portions that require restoration, as in VFT 1 and VFT 5. Figure 7.2 shows another refuge reference site.

In instances where an auto-reference or refuge is available, the reference sometimes serves almost as if it were a template for ecosystem recovery, particularly if the degraded ecosystem requires little intervention to return it to its former, intact state. Consider a grassland or savanna that has been degraded by colonization of woody plants following an extended period of fire suppression. Its original plant species composition may persist for an extended period in dormant condition as a *seed bank* or propagule bank. All that would be needed is to ignite several fires with short return intervals during growing seasons to kill invasive shrubs and young trees and to remove an accumulation of leaf litter and coarse detritus. Thereafter, the degraded ecosystem will restore itself to a condition that cannot be distinguished from its preimpairment state. The restored ecosystem in figure 1.1 serves as an example. An auto-reference may also serve as a template in more severely damaged ecosystems that consist of only a few species and a predictable structure, such as subtropical mangrove forest.

More commonly, a reference is not a template but rather an imperfect vision

**FIGURE 7.2.** Monkey-puzzle tree (*Araucaria araucana*) forest in Chile that serves as one of several reference sites for planning restoration on lands that had been converted to *Eucalyptus* plantations.

to be approximated, hence our use of the terms "beacon" and "pointer to the future." This circumstance applies to reference types (3) and (4) of White and Walker (1997). In type (3)—different time, same place—reference information is available that characterizes the ecosystem before its decline or demise. Such information may include descriptions and photographs of the ecosystem prior to its impairment, or historical documents may describe the natural history of that locale. In contrast, in type (4)—different time, different place—essentially no information is available on prior ecosystem conditions at the project site, but such information is available for one or more regional ecosystems of the same kind that occupy a similar landscape position with similar physical site conditions.

Sometimes no reference sites are available for study, or they are remnants that are incomplete and not entirely adequate, in which case the reference model must be assembled, at least in part, from secondary sources of evidences. Figure 7.3 shows a helpful but incomplete reference site for a restoration project north of Sydney, Australia, and figure 7.4 shows a historical document that supplies secondary reference information to supplement information gleaned from the woodland shown in figure 7.3. In regions where reference sites are unavailable,

**FIGURE 7.3.** Small remnant of original forest that served for the preparation of the reference model for the Kooragang Island restoration project site, New South Wales, Australia. This is a same-time, same-place reference site.

and if the intent of ecological restoration is to restore an original ecosystem rather than a more recent semicultural ecosystem, then data from palynological and archaeological records can be accessed and interpreted with caution. Much pollen can only be identified to genus or family, and its relevance is limited. Windborne pollen did not necessarily originate in the site where it was deposited, and insect-pollinated species may be underrepresented or entirely absent from fossil pollen samples. Tree ring analysis and carbon dating of identifiable plant parts preserved in peat bogs or lake sediments can reveal much about original systems.

More recent sources of secondary evidence can be helpful to develop reference models for restoring semicultural ecosystems and also original ecosystems that were obliterated relatively recently. These sources include lists of native species from published floras and faunas and specimens deposited in herbaria and museums. Patient detective work can provide scraps of evidence from examinations of historical documents, historical photographs, diaries, and examinations of old accessions in botanical gardens. Art museums can impart a wealth of information from old landscape paintings. Even the botanical contents of preserved packrat middens, can provide paleoecological evidence (Rhode 2001).

**FIGURE 7.4.** Historical document from the mid-nineteenth century describing species known to occur at the Koorangang Island (NSW, Australia) restoration project site at that time. This is a secondary source of evidence for the preparation of the reference model for this restoration project.

## Steps in Reference Model Preparation

How does one go about preparing a reference model for a new ecological restoration project? Clewell and McDonald (2009) identified several steps in the preparation process. These steps are presented below with a few refinements added. The steps assume that the vision of the restored ecosystem has already been formulated and that reference sites were already designated. In addition, the goals of restoration have been enunciated. Prior to the finalization of the reference model, and preferably before it is drafted, the impaired ecosystem will have been inventoried to discover what ecological legacy yet remains. This legacy, in terms of species, biotic structure, and abiotic structure that survived impairment, form the core around which restoration tasks are developed. The degree to which the ecological reference can serve as a model for a restoration project depends on its content, which varies widely among projects. In some projects the reference model can serve almost as a template. In others, reference information is scant, and it can only hint at the direction of development.

## Assemble Documentation Needed to Prepare the Reference Model

Documents that provide ecological descriptions of reference sites need to be identified and accessioned for use, preferably in digital format. If available, these will include aerial photographs; maps that show geographic, topographic, and soils data; and technical reports and publications that ecologically describe reference sites. If these ecological descriptions are not available, field studies may have to be conducted in order to gather the information that will be needed for preparing the reference model. If secondary evidences are used in preparing the reference model, a search for documentation may involve wide-ranging scholarship and detective work. Those engaged in such searches should be familiar with *The Historical Ecology Handbook* (Egan and Howell 2001).

A single reference site may be too small for a technical description that includes a full range of species and other biophysical properties that occur locally in that kind of ecosystem and that could appear at a project site during the course of ecological restoration. A restoration cannot be considered unsatisfactory because it contained locally characteristic species that were not recorded in the inventory of a single reference site. To the extent possible, a reference model should be prepared that reflects the gamut of potential variability in biotic expression in terms of species composition and community structure. Therefore, the reference model serves best if it is a composite of the species composition and other biophysical characteristics of more than one local ecosystem of the same type, landscape position, and general site characteristics as the impaired ecosystem undergoing restoration. A restoration project should be considered successful if biotic expression falls within a range of variability that was determined from multiple reference sites (Baird and Rieger 1989; Clewell and Lea 1990). Reference models based on ecological inventories of target ecosystems can be enriched by whatever secondary evidence may be available.

## Prepare Documentation

This step consists of synthesizing a reference model from all of the available information. This document does not have to be a formal report as if it were to be published or widely distributed. It is written primarily for the benefit of the restoration planner and others directly involved in the project. It only needs to contain information that the planner requires to prepare planning documents.

If, for example, the project site has not been impaired with respect to hydrology, then hydrology would not have to be considered in the reference model, except briefly and generically. The reference model does not even have to be a single document. Instead, it can be an annotated index to preexisting sources of documentation that collectively provide the information needed for the preparation of project plans and for the project director to approve those plans.

The degree of detail needed in the reference model will reflect the degree of impairment and thus the intensity of the restoration effort. A restoration project that only requires the removal of some patches of harmful invasive organisms, for example, or the resumption of prescribed fire at a frequent recurrence interval, the reference model can be brief and concentrate only on those few aspects of the biophysical environment that need attention. Any additional effort put into the preparation of the reference model would unnecessarily absorb time and funds. If the restoration project requires reconstruction of the entire ecosystem following its destruction during surface mining, the reference model should be much more comprehensive and detailed, and it may warrant the attention that would be given to preparing an ecological monograph (e.g., Clewell et al. 1982).

Plant species composition is frequently the single most important inclusion in a reference model, particularly if desirable species must be introduced or undesirable species removed. If environmental conditions are changing, the species composition of the reference model can be augmented with several characteristic local native species that prefer the anticipated new conditions. For example, if the climate is becoming drier, the reference model for the prior state can be augmented by plant species that ordinarily occur locally in drier communities. Few individuals of those species need be introduced at the restoration site—only enough to reproduce and populate the ecosystem if drier climatic conditions indeed occur. The species selected for introduction should be those that would be expected to colonize independently of restoration, if drier conditions occurred and as long as seed sources for them were available in the vicinity. The selection of other species that were uncharacteristic of the local area would be tantamount to *rehabilitation* (chap. 10), not restoration.

For example, let us consider a situation where a cypress-blackgum swamp (*Taxodium ascendens* and *Nyssa biflora*) was being restored in central Florida, but widespread agricultural irrigation was depleting groundwater and lowering the water table to the point that swamps might not be able to support cypress and blackgum much longer. The practitioner could plant a few sweetbays (*Magnolia virginiana*) and loblolly bays (*Gordonia lasianthus*) on the most elevated terrain at the project site, where they could persist and eventually serve as seed trees to gradually replace cypress and blackgum if groundwater depletion continued. If conditions had already changed, then the project site should be planted with sweetbays and loblolly bays to the exclusion of cypress and blackgum. The reference model should be flexible to allow these options in light of site specific conditions.

For all projects, the content of the reference model needs to be sufficient for project plans to address every stated project goal. For example, if a goal of restoration is to provide habitat for a rare, animal-pollinated plant species, the reference model would benefit by specifying its pollinator for possible reintroduction and any special habitat needs that it may require.

## Identify Anomalies

Most reference sites that are inventoried for the preparation of a reference model contain a few alien species or native nuisance species (chap. 3). Moreover, in spite of their overall high quality as natural areas, these sites commonly display human-mediated environmental scars, such as roadbeds or other intrusions of infrastructure and disturbance. Such species and physical features are undesirable for inclusion at a restoration project site. We refer to them as anomalies and delete them when preparing a reference model, or we specifically flag them so they will not be incorporated into project plans. Similarly, anomalous species could be listed among those that can be considered for removal if they appear. It would make no sense to restore an ecosystem to a degraded condition or to penalize a restoration for not emulating a degraded condition.

## Identify Critical Elements

The reference model should prioritize those ecological elements of greatest concern. These elements may be the reintroduction of *keystone species* or the establishment of a critical functional group. Alternatively, the emphasis may be placed on abiotic elements such as returning the water table to a critical elevation or performing site preparation in a particular manner. For example, site preparation and seed planting for VFT 1 consisted of several tasks, all of which needed to be executed with precision to guarantee the development of desired ground cover vegetation. The reference model should emphasize the importance of such attention to detail.

Most restoration projects for terrestrial systems concentrate on reestablishing vascular plants to ensure recovery of community structure and those species responsible for primary production. In many instances, animal populations will eventually colonize a restored ecosystem on account of their mobility. The reference model can identify any species of animals that are not likely to return to the restored ecosystem spontaneously, so that project plans can specify their reintroduction. Ecologically important cryptic species should be designated for reintroduction as needed, such as mycorrhizal fungi, nitrogen-fixing microorganisms, other soil biota, and benthic organisms.

Community structure commonly reassembles itself naturally, as long as species composition and the physical environment are suitable. If any elements of community structure will require the attention of a practitioner, they should be identified specifically in the reference model, so that project plans can be prepared accordingly. In grass-dominated ecosystems that require frequent fires to maintain structure, the reference model can emphasize the importance of flammable groundcover, so that plans can be prepared for the rapid establishment of these fine fuels.

The identification of environmental drivers and stressors that maintain ecosystem integrity is an essential component of reference models. For example, fire regimes are identified for pyrogenic ecosystems, and hydroperiods are described for wetlands. Project plans and prescriptions for environmental stewardship and postproject management will be inadequate without satisfactorily addressing the dynamics of environmental drivers.

## Identify Missing Elements

In some ecosystems, characteristic species cannot be recovered because of recent extinction, local extirpation, or the unavailability of seeds and planting stocks. For example, restoring forests in the Appalachian Mountains becomes problematic on account of the reoccurrence of the blight (*Cryphonectria parasitica*) that has driven the formerly dominant American chestnut (*Castanea dentata*) to the brink of extinction. Important tree species in historic times have been depleted by repeated logging to the point that their importance, or even their occurrence, may be overlooked in preparing reference models. Depleted species need to be noted in the reference model, so that they are included in restoration plans. Characteristic herbaceous species in forests may have been extirpated over broad areas by feral pigs and will not appear in inventories to develop restoration models without botanical sleuthing.

Restoration plans can identify substitutions for species that cannot be recovered, especially if replacement species can provide an ecological function previously provided by missing species in the community. These species should not be selected from a seed catalog. They should be coadapted species and belong to plant communities associated with the biome in question. Replacement species can be suggested in the reference model, if they are not already treated elsewhere in project plans. In either event, the nomination of substitute species requires an evaluation of the potential to fulfill ecosystem services without compromising biodiversity that contributes to the historic continuity of the ecological trajectory.

## Temporal References

Most references are selected because they represent mature biotic expressions of the kind of ecosystem designated for restoration. A recently restored ecosystem, though, commonly represents a younger stage in ecosystem development, particularly if the project site had been largely cleared of vegetation during impairment or site preparation. In evaluating restoration success, a comparison between systems of contrasting ecological stages requires interpretation to overcome that discrepancy. Early seral species may dominate a recently completed project site. Most of these species may be absent at reference sites. Species abundance will certainly change as a system matures. The comparison would be much more valid if the reference site represented the same ecological stage of succession as that of the restored ecosystem. Significant and inescapable subjectivity is introduced into project evaluation when a newly restored ecosystem of a decidedly immature seral stage is compared with a mature reference model. To overcome this problem, several restorationists have compared ecosystems undergoing restoration to equivalently immature ecosystems that were recovering by natural regeneration.

The first such attempt was reported by Feiertag et al. (1989) with respect to the restoration of sand pine scrub (*Pinus clausa*) and ecologically similar scrubby pine flatwoods in Florida. Both communities are fire-maintained and occupy internally well-drained crests of low sandy ridges. They are sometimes restored on land that was physically recontoured by backfilling following opencast mining for phosphate ore. Restoration is commonly conducted by mechanically salvaging topsoil that contains rootstocks and seed from a site about to be mined and spreading it on mined and physically recontoured land nearby. Feiertag et al. (1989) prepared a reference site by removing topsoil from a previously undisturbed site and immediately respreading it on the same site. Unlike mined and recontoured land, the reference site had suffered no disruption to its deep soil structure and hydrological regime from mining and was allowed to recover by unassisted biological succession. Recovering vegetation was monitored two years later at both restoration and reference sites. Similarities in composition suggested recovery along the same ecological trajectory for both reference and mined sites.

Grant (2006) applied a variation of this approach for evaluating the restoration of jarrah forests (*Eucalyptus marginata*) in Western Australia, replanted on lands that had been backfilled and physically recontoured after opencast mining for bauxite. Elaborate monitoring of both vegetation and physical parameters was conducted regularly on restored lands of various ages since restoration began. Restored jarrah forests on the oldest recontoured sites had developed sufficiently to determine that this restoration strategy was indeed satisfactory for the recovery of the premining ecosystem. Therefore, monitoring data representing earlier stages of restoration were used as sequential reference models for newly restored sites on more recently mined and recontoured lands. For example, a new, two-year-old restoration site would be compared to former two-year-old restoration sites. If monitoring values for important parameters at the new site fell within the range of values for those same parameters at former, and ultimately successful, restoration sites, then the new restoration was considered satisfactory. If those values fell outside this range, then midcourse corrections were to be conducted. In this example, previously restored project sites become sequential reference models for newly restored project sites.

Similarly, reference states in various degrees of degradation can be used as reference models for an ecosystem in sequential states of recovery, as illustrated in figure 7.5. The main message of this figure is that as ecosystems become overexploited or degraded in regard to one particular ecosystem service that it provides, various species and one or more ecological processes in which they participate "drop off" (left, top to bottom) and ecosystem health, integrity, and resilience decline. The flow of one or more ecosystem services may increase short term, but the services that the ecosystem as a whole provides will inevitably decline. The distortion and diminution of the circles representing an ecosystem within its biophysical and socioeconomic matrices is further indication of transformation that represents damage and degradation. Attempts by practitioners to repair may prove difficult, because ecosystem parts and processes may not fully emulate the reference system on account of irreparable damage to the environment or intensive land use. In such instances, multiple references in different stages of a restoration project provide a reasonable and more realistic way to measure restoration progress over time. A further point to consider is that constructed or *designer ecosystems* can also be compared to both unimpaired reference systems and systems undergoing restoration. The flow of one or more ecosystem services provides an excellent axis for comparison across a range of situations of this kind, as shown recently for a large selection of restored and created wetlands by Moreno-Mateos et al. (2012).

## References in Landscape Restoration

In landscape-scale restoration programs, ecological restoration may be conducted in concert with the rehabilitation of degraded production systems and other natural and seminatural areas in the same landscape, in order to promote long-term ecological and economic sustainability. Landscape-scale programs that are integrated with investments in social capital with the intent of improving the well-being of affected communities are known as RNC programs (restoration of natural capital, chap. 10). RNC programs may cover large geographic areas

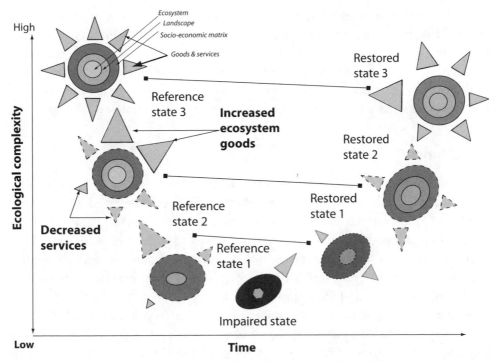

**FIGURE 7.5.** Sequential references in ecological restoration. Dashed lines represent degraded or fragmented conditions as compared to a whole system and integrated landscapes. The inner circle represents the ecosystem. The one or two outer circles represent the landscape and the socio-economic matrix in which the restored ecosystem is embedded. The triangular appendages represent the various natural goods and services that accrue from an ecosystem.

and may be conducted sequentially over many years. Multiple reference models may be needed for the different kinds of impaired ecosystems in need of restoration or rehabilitation. In socioeconomic landscapes that have been inhabited continuously for millennia, the reference models will necessarily reflect cultural influences, as described in chapter 6. This gives stakeholders and sponsors of restoration programs flexibility regarding the time period and its environmental conditions that serve as the ecological reference. During the course of a landscape restoration program, the restoration model can be adjusted to reflect different time periods.

Reasons to make adjustments are to improve restoration outcomes relative to those obtained earlier in the program and to accommodate changing desires of stakeholders. A strategy to allow such adjustments to the reference model in a manner that ensures ecological and economic sustainability—called Historical Multi-Criteria Analysis (HMCA)—was devised by Aronson et al. (2012). It is intended to facilitate consensus in choosing sequential ecological references that parallel the historical stages in degradation and transformation of a given ecosystem or landscape. The HMCA strategy was adopted in a coastal lagoon restoration program near the Mediterranean coast of France. This approach has helped some of the major stakeholders and financiers to prepare a clearer, long-term vision of their shared values and goals for this diverse and heavily populated landscape.

## Trajectories

As we saw in chapter 1, an ecological trajectory is the sequence of biotic expressions of an individual ecosystem over time. It could be conceived as a moving picture of an ecosystem that covered centuries or several millennia and that was speeded up so that its development could be viewed in less than an hour. That movie would show the continuous adjustments that a dynamic ecosystem undergoes, such as changes in composition of its conspicuous species and changes in community structure and life form frequencies.

Unlike an ecosystem itself, an ecological trajectory is not real or tangible. Instead, it is a progressive historical record of what an ecosystem was like. A trajectory can be projected into the future, as long as future environmental conditions can be ascertained with a reasonable degree of confidence and the network of interactions among the ecosystem and its neighboring systems remains constant. If no changes in future environmental conditions are anticipated, then the ecosystem is not expected to change much in the future. If an ecosystem were impaired during such a "calm" period of environmental stability, it could be restored with a high degree of *historic fidelity* to its former state. The analogy of a seagoing vessel that ran afoul on rocky shoals (chap. 3) applies on the assumption that the vessel will continue on its previous course following rescue and repair.

However, change is constant in Nature, although its rate varies. When the rate of change is rapid, an ecosystem, as reflected in its trajectory, can change quickly. For example, grassland that occupied cold steppe shortly after the retreat of Pleistocene glaciers could change to coniferous woodland and eventually to mixed deciduous forest in response to a warming climate. This trajectory could be elaborated by cycles of natural disturbance that return the ecosystem to an "immature"" or seral stage of development from the perspective of succession theory.

Sometimes we want to restore an impaired ecosystem to a seral state and then manage it to keep it in that state of interrupted succession, as in European chalk meadows (chap. 4), California oak savannas (chap. 6), and Ohio oak-hickory forests (chap. 6). From that perspective, we must sometimes consider ecological references as dynamic systems that change over time. In other words, we need to add ecological trajectories to the consideration of ecological references, so that the expressions of ecosystems that we restore and later manage represent stages in natural trajectories rather than designer landscapes.

However, there are several problems with applying the notion of ecological trajectories too strictly. First, the original ecosystem may not be available as a reference site, and the trajectory will have to be assumed on the basis of substitute ecosystems. In those cases there is always a degree of subjective choice, judgment, or negotiation (Aronson et al. 1995). Second, the environment may change and cause a shift in trajectory. Third, patch dynamics (chap. 4) may cause an unanticipated change in an ecosystem and thus its trajectory.

A restoration practitioner initiates ecosystem processes but does not control subsequent ecosystem dynamics. Planting certain species at particular densities may effectively nudge a recovering ecosystem toward a desired biotic expression but does not guarantee its attainment. Spontaneous colonization by other species and their competitive exclusion of planted species can radically change a trajectory. The option remains to continue manipulating species populations indefinitely. However, such a practice can no longer be called ecological restoration. Instead, it is a form of management, gardening, or engineering, whereby an intentional design is imposed upon a natural system. The result is human artifice rather than recovery of Nature. In such instances, products in terms of target states are valued more than the emergence of ecological attributes (chap. 5).

Above all, a trajectory should not be used as a norm for evaluating the success of an ecological restoration project. Too many environmental, biological and ecological variables exist for a trajectory to have rigorous predictive value. The same admonition applies to the reference model. Both the reference model and its anticipated trajectory should be considered as points of departure rather than endpoints.

Chapter 8

# *Approaches to Restoration*

In ecological restoration project work, we assist the recovery of impaired ecosystems (SER 2004). In chapter 4 we raised the question concerning how much assistance we should apply to recover an impaired ecosystem. This chapter explores that question at some length. Thereafter, we introduce the framework approach to restoration, and the chapter concludes by examining the sources of knowledge used in ecological restoration and how these sources influence our approaches to performing restoration.

## Intensity of Effort

What level of effort is needed to accomplish ecological restoration? As with everything else in ecological restoration, the answer is site specific; there is no rule or recipe that fits all sites. We approach this topic by describing four levels or intensities of effort that can be applied or attempted, levels borrowed from similar schemes for this purpose that were proposed by T. McDonald (2000) and by Prach et al. (2007):

1. Prescribed natural regeneration
2. Assisted natural regeneration
3. Partial reconstruction
4. Complete reconstruction

### *Prescribed Natural Regeneration*

Prescribed natural regeneration is ecological restoration in which project tasks include no biophysical manipulations or other direct interventions at the project site or in its surrounding landscape. Instead, ecological recovery relies on natural regeneration, which, in most instances, consists of what is commonly called plant succession (Prach et al. 2001). The principal interventions in prescribed natural regeneration projects consist of removing the sources of disturbance that cause impairment and protecting the project site sufficiently for natural recovery processes to occur. Like all ecological restoration, prescribed natural regeneration requires prior intent and planning before the project begins, or else it would not be prescribed. Furthermore, without prior intent there could be no assistance, as required by the definition of ecological restoration (SER 2004) in the phrase "assisting the recovery of an ecosystem." Spontaneous recovery that occurs without prior intent fails to meet the definition of ecological restoration. Prescribed natural regeneration requires prior knowledge that successful recovery is likely to occur without resorting to biophysical manipulations and that recovery will reestablish historic continuity as predicated by reference criteria.

Prescribed natural regeneration is the least intrusive and least costly level of intervention for accomplishing ecological restoration. Some authors use the term *passive restoration* in place of prescribed natural regeneration. We eschew that practice, since ecological restoration by definition is not passive, as we already noted in chapter 4. Moreover, passive restoration is commonly used for indicating natural recovery, which lacks prior intent, and the term therefore merits rejection as a source of confusion.

The most celebrated example of prescribed natural regeneration is tropical dry forest restoration on 110,000 hectares of marginal farmland in the Guanacaste region of Costa Rica by Daniel Janzen. Restoration consisted of purchasing the land and then letting "nature take back its original terrain" (Janzen 2002, 559). Although intentional tree planting and soil conditioning were attempted in limited areas, these manipulations were subsequently halted and most terrain within the large project area underwent no direct manipulation of any sort. Premeditated interventions that distinguished the Guanacaste project as ecological restoration were the acts of protecting land from ranching, anthropogenic fire, subsistence agriculture, irrigation, hunting, and sporadic logging. Prior ecological investigations demonstrated that spontaneous forest reseeding would follow. Ecological inventories of the few relatively undisturbed tropical dry forest that survive as remnants of a formerly vast and continuous forest covering much of Central America, represented a reference model. These same forests served as sources for natural seed dispersal. Janzen had already determined that trees with wind-dispersed seeds would replace pasture grasses if protected from grazing and fire, and that trees with animal-dispersed seeds would follow, once a forest canopy had developed. Therefore, the Guanacaste project was initiated with advanced knowledge of how it would occur and with confidence that it would reestablish historic continuity.

Other well-known, carefully studied examples of natural regeneration come from central Europe, where plant communities have regenerated without assistance on former mined lands (Prach and Pyšek 1994; Rehounková and Prach 2008). At some of these sites, regeneration was intentional and could be counted as ecological restoration, because natural regeneration was predicated on knowledge of seed dispersal from nearby natural forests that served as reference sites and from knowledge that a substrate suitable for soil formation had been prepared during physical mine *reclamation*.

## Assisted Natural Regeneration

Assisted natural regeneration describes ecological restoration projects in which the impacts from impairment are eliminated by nonintrusive biophysical manipulations that are more subtle than overt. Assisted natural regeneration is applicable where the physical environment, if impaired, can be repaired with minimal effort in small areas. Assisted natural regeneration releases desirable native species from competition without applications of agrichemicals and mechanized equipment, at least over broad areas. Common tactics include prescribed burning, nucleation plantings, deposition of debris mounds, building of micro catchments to capture water and nutrients, similar measures to promote niche diversity, and placement of perches for frugivorous birds. These interventions represent minimal subsidies with high returns by removing obstacles to ecological recovery (M. C. McDonald 1996; Clewell and McDonald 2009).

Shono et al. (2007) provided an example of assisted natural regeneration from the Philippines, where practitioners stood on boards placed over dense swards of invasive cogon grass (*Imperata cylindrica*), thereby crushing it and halting its growth. This in turn facilitated the establishment of saplings of native forest trees (dipterocarps), whose seedlings were released from competition that would otherwise have killed them. The process was repeated approximately three times a year for two or three years before the trees were large enough to overcome competition without further assistance. No herbicides were used. Thereafter, young trees were sufficiently established for their crowns to begin coalescing into a canopy that further dampened cogon grass competition.

Nucleation plantings—sometimes call pocket plantings—are small, densely planted areas, usually with at least several species, often distributed like stepping stones of varying sizes (Holl et al. 2012). If the planting consists of a grove of trees, they are called tree islands. They serve as points of radiation for the spread of desirable species into adjacent, unplanted areas. They serve as microsites where favorable soil conditions develop and from which the soil biota can disperse in a centrifugal manner as the organic matter content in the soil improves in surrounding areas (Cole et al. 2010). They attract birds, bats, and other dispersers of seeds and fruits of native plants from nearby forest remnants. In harsh environments, nucleation plantings are placed in relatively protected microhabitats called *safe sites*, as was done experimentally to restore degraded ski slopes in the Alps (Urbanska 1997) and tropical montane forest in Central America (Zahawi and Augspurger 2006). Nucleation sites can also serve as *infiltration areas* where spontaneous dispersal and establishment is favored. Tree and shrub nucleation sites attract bird-dispersed plant species with fleshy fruits, including alien species (Milton et al. 2007).

In arid, windy sites, debris mounds have been successfully installed to capture drifting seeds and detritus (Tongway and Ludwig 2011). Trapped debris becomes incorporated as organic matter in the soil, and seeds germinate and provide vegetative cover on previously exposed sites. Microcatchments are typically used in arid land restoration and consist of small, V-shaped or crescent-shaped berms one or two meters long and upwards to thirty centimeters tall that trap surface runoff and detritus to stimulate plant growth (Bainbridge 2007).

Bird perches have been constructed from posts with a few crossbars attached and placed in open areas designated for restoration. Birds that feed in nearby forests are attracted to the perches. Undigested seeds that pass through bird guts inoculate exposed soil beneath with desirable plant species. In Australia, snags of poisoned, exotic trees are purposely left standing for the same purpose (McDonald 2000). For this technique to be successful, seedbed conditions must be favorable and preexisting vegetation not competitive if germination and seedling establishment is to occur (Holl et al. 2000).

John Tobe (personal communication) succeeded in assisting the natural regeneration of upland hardwood forest by interplanting seeds and freshly extracted transplants from local donor sites beneath the partial shade of a young pine forest that volunteered on abandoned agricultural land in Georgia, USA (figure 8.1). The restoration was accomplished quickly and at virtually no cost, although the restored area was small. The effort greatly accelerated natural succession and resulted in characteristic species of both trees and undergrowth that ordinarily fail to appear in secondary forests in that region.

## Partial Reconstruction

Projects designated as partial reconstructions rely partly on technical solutions and partly on natural regeneration. Technical solutions may include the mechanized repair of the physical environment using civil engineering methods. For example, stream banks may be reshaped and ditches filled. Other technical solutions rely on agronomic tactics such as subsoil ripping and disk harrowing, the broad-scale

**FIGURE 8.1.** Mesic hardwood forest restoration performed by interplanting trees and undergrowth species in shade beneath a stand of young pines in Georgia, USA. Pines are removed gradually as hardwood trees attain height and cover.

application of agrichemicals, mulch application, mechanical seed dispersal, and the mechanized outplanting of nursery-grown stocks. These methods are applied to wide areas as opposed to manually installed nucleation plantings.

Reconstruction of this sort becomes necessary when natural resilience is depleted and biophysical elements of the ecosystem need replacement before recovery can be expected within acceptable time limits. Prach et al. (2007) explained that technical solutions sometimes can be implemented by the sequential removal of barriers to species establishment. Initially, removal of these barriers filter out impediments to the recovery posed by the physical environment or to the dispersal of species. As recovery proceeds, barriers in terms of competition and other biotic interactions are removed. Technical solutions attempt to remove all of these barriers. It is to be noted, however, that a recent survey of over 1,500 wetland restoration projects showed that restorationists sometimes apply more technical effort at a site than needed (Moreno-Mateos and Aronson, in review). This practice increases costs and threatens ecological effectiveness. A lesson imbedded here is that sometimes less is more, and it needs to be considered on a site-specific basis.

Virtual field trips describing partial reconstructions include shola grassland restoration (VFT 3), Brazilian forest restoration (VFT 7), restoration of flood plain (but not the stream) in Oregon (VFT 6), and temperate evergreen rainforest restoration in Chile (VFT 7). Most partial reconstruction projects involve growing stock in nurseries for outplanting.

### Complete Reconstruction

In complete reconstruction, all phases of recovery are characterized by manipulations of the biophysical environment. Projects depend entirely on technical solutions rather than on natural regeneration. Some natural regeneration may occur but not to a substantial degree or not that can be anticipated with confidence by project planners. Complete reconstruction is sometimes the only option available when the ecosystem to be restored has been entirely destroyed. The most obvious example is restoration conducted on a site that was surface mined and the mine pit backfilled with overburden. Surface materials are usually reworked and amended to facilitate soil formation. Hydrological conditions may require considerable adjustment.

Mines commonly occur in denuded landscapes, where natural dissemination of plant propagules is unlikely, except for ruderal species. The widespread outplanting of nursery-grown stock is generally needed to introduce desirable species and control erosion. The species that are outplanted and their initial abundance will largely determine the outcome of restoration. For example, outplanted trees will characterize the first forest generation, which may persist for at least several decades. Competition and in situ production of seeds from these trees may overwhelm colonization by other species, and the effects of outplanting could influence forest diversity in subsequent generations.

Among the virtual field trips, longleaf pine savanna restoration (VFT 1) was a complete reconstruction, and so was the restoration of Mediterranean steppe (VFT 2). Stream channel restoration in Oregon (VFT 6) represented complete reconstruction.

Prescribed natural regeneration, assisted recovery, and reconstruction are not mutually exclusive approaches. They are best considered nodes along a continuum that describe the intensity of intervention. Subtropical thicket restoration (VFT 4) falls between assisted natural regeneration and partial reconstruction. The point here is to recognize that options exist with regard to intensity of restoration effort and its consequences in terms of project quality, project cost, and the time to project completion.

At project sites that require reintroductions of numerous kinds of plants, it may be sufficient—and certainly less costly—to introduce relatively few individuals of many plant species than to plant large numbers of each. If site conditions are suitable for a species, and several individuals are planted with care, they will grow with vigor, reproduce, and increase their abundance without assistance. This technique represents assisted natural regeneration and was applied to reestablish forest undergrowth during headwater wetland forest restoration on surface-mined and physically reclaimed land (Clewell 1999). Several plants each of ~ thirty species of undergrowth herbs and a few shrubs were removed from natural forest and transplanted into a thicket of previously planted tree saplings at a restoration project site. After two years, many transplants representing most but not all of these species had increased their numbers substantially. Some became abundant in the newly restored forest. The preferred strategy is to overintroduce species rather than to overplant individuals of a given species (Brancalion et al., in review). It hardly matters how many individuals of poorly adapted species are planted if none survive.

We return now to the question we posed earlier: what should be the intensity of restoration effort? Should we merely nudge natural processes and let nature take its course, or should we fully superimpose technical solutions? McDonald (2000) emphasized that complete reconstruction requires high investment in financial and human resources. She argued that complete reconstruction overwhelms natural recovery and imposes a predetermined outcome rather than stimulating natural processes whose outcomes are less predictable. In contrast, McDonald (2000) claimed that working with natural recovery processes can produce a restoration outcome with a high degree of fidelity to the predisturbance state, "because it builds bridges for the continuity of often complex and irreplaceable components." She insisted that nothing is gained from the excessive actions that characterize complete reconstructions. Sampaio et al. (2007) arrived at the same conclusion after finding that intensive restoration efforts may slow recovery of forest in Brazilian pastures. Although this advice generally applies, exceptions exist, as demonstrated in VFT 1 for xeric longleaf pine savanna, where complete reconstruction is the only option for ecological restoration.

The greater our intensity of intervention, the quicker—in theory—the ecosystem we target for restoration can recover to the point that it no longer needs our assistance. However, project costs increase with intensity of effort. Restoration projects that are completed with high intensity of effort may show signs of artifice, such as row-planted trees or riverbanks covered with riprap. Rapid completion allows little opportunity for aftercare following initial project implementation and limits the time practitioners have to make midcourse corrections and otherwise address unanticipated threats to project success.

The selected approach may depend on a variety of factors and influences, including these:

1. budget constraints (prescribed natural regeneration is the least expensive option);
2. time limitations (reconstruction is accomplished in the least time);

3. availability of labor and equipment (prescribed natural regeneration requires the least of each);
4. specified project goals (goals, other than attainment of biophysical attributes of a restored ecosystem, may require intensive interventions);
5. level of disturbance (repair of the physical environment requires more effort than manipulations of the biota);
6. landscape context (intensive interventions may be needed to compensate for reduced exchanges of organisms and materials in urbanized and highly impaired landscapes);
7. availability of technological options (research and development for new methods will increase costs);
8. contractual, legal, and administrative realities (permit conditions may preclude some restoration options);
9. political realities (restoration tasks that require prescribed fire may be forbidden on account of political pressures);
10. ecological issues (e.g., there may be need for rapid, intensive implementation in order to deter colonization of open habitat by invasive species, or to conduct planting during brief periods on account of restrictions of weather or climate);
11. socioeconomic priorities (the need to accelerate restoration to retain flood water, prevent erosion, or provide another ecosystem service).

If none of these issues is applicable, the preferred level of intensity is to make only those manipulations that are needed to reengage ecological processes in a manner that relieves the need for human subsidy thereafter. This strategy was advocated in the 1960s by two sisters from Australia, Joan and Eileen Bradley (Bradley 1971), and it continues to serve as the guiding principle for the restoration of natural areas on public lands in Australia today. This policy warrants emulation elsewhere and deserves consideration whenever latitude exists to select the intensity of restoration effort.

The attainment of this low-input ideal is largely restricted to community projects and to local stewardship programs with long timeframes. Other restoration projects, which provide the principle employment opportunities for restoration practitioners, are driven by the missions and policies of large organizations that sponsor and regulate restoration projects. The common rationale for restoration in these larger projects is to provide ecosystem services, particularly substrate stabilization and erosion control; detention of flood waters; improvement of water quality and water storage; habitat for officially protected species; habitat for other desirable wildlife; and recreational opportunities associated with natural areas. These are usually accelerated projects with short-term *objectives* to provide these ecosystem services, and they are not intended specifically to reestablish historic continuity and satisfy the ecological attributes discussed in chapter 5.

## Framework Approach

In chapter 4, we raised the issue of accelerating ecosystem recovery by reintroducing all desirable species at one time in an attempt to eliminate plant succession at a restoration project site or at least greatly reduce the time needed for succession until ecological processes had returned to normal rates of functionality. This tactic was formalized by Goosem and Tucker (1995), who proposed the framework approach while working in Queensland, Australia. The approach was later attempted in northern Thailand (Elliott et al. 2003) and in Brazil (Souza and Batista 2004). Typically, twenty to thirty tree species that represented the main successional stages of forest ecosystems were planted. These trees included fast-growing, early-seral trees with large crowns that were planted together with slow-growing and narrow-crowned, mid- and late-seral species in plantings that covered the entire restoration site (Rodrigues et al. 2010). The reestablishment of a multilayered forest canopy rapidly suppressed heliophilous (sun-loving) weeds, including invasive grasses and ferns. This canopy promoted a weed-free understory and facilitated natural regeneration of unplanted native species, while stimulating the dynamics of nutrient, water, and energy cycles (Wydhayagarn et al. 2009). Other techniques for this approach in tropical forests are under development in Costa Rica, Brazil, Indonesia, and elsewhere, as described in Lamb et al. (2005) and Holl (2012).

The framework approach seems best suited to those regions of the world where competitor species (Grime 1974; chap. 3) are prevalent. Where stress tolerators prevail, practitioners routinely plant species typical of ecologically mature forests on denuded sites (e.g., Clewell 1999), under the assumption that the process of initial floristic composition (chap. 4) is applicable.

## Sources of Knowledge

Epistemology—the study of knowledge and its origins—has never been a popular undergraduate major. The sources of knowledge in ecological restoration are divergent, interesting, and largely misunderstood. We attempt to illuminate these sources and show how they have influenced our approaches to restoration.

### Intelligent Tinkering and Knowledge-based Restoration

Ted Sperry pondered an abandoned agricultural field on the outskirts of Madison, Wisconsin. It was 1934. He had just been engaged to transform that field into tallgrass prairie, the way it once looked in the nineteenth century. He had grown up in farmland and had recently received a university degree in botany. He knew something about the vegetation and natural history of prairie remnants that had escaped the plow along railroad rights-of-ways and in cemeteries. He had access to a farm truck and a crew of able-bodied men armed with shovels from the Civilian Conservation Corps at his disposal. Put yourself in his shoes. How would you have figured out what to do first? Would you plow the field in preparation to sow a cohort of native species? Would you also spread manure? Where would you find seeds? What would you do about all of those agricultural weeds that would certainly compete fiercely with seedlings from native prairie plants? How would you cope with trees that would seed-in from surrounding forests? Was there another way to begin? The assignment did not come with a book of instructions. Ted had to come up with something on his own and give it a try.

The project, now known as Curtis Prairie, turned out quite well, although orchard grass and a few other weeds of European origin were never quite eliminated. It was one of several prairie restoration projects that quietly began at about the same time in the upper Midwest and spread during the next twenty years across the United States, as people tried to recover other kinds of ecosystems. A comparable movement developed in Australia in the 1930s, building on the work of Ambrose Crawford and Albert Morris (T. McDonald 2008). All of the practitioners of these projects faced the same problem. What do I do first? Then what comes next? There was no place to turn for help. Practitioners designed their own projects; there were no professional planners involved. The only option was to dig into one's knowledge of natural history and one's experience as a farmer or gardener and get to work. If one thing didn't work, you tried something else, but not without forethought. You learned from your mistakes. You thought back to what you learned in college and how you helped your dad plant seeds on freshly plowed land.

Aldo Leopold (1949) had a name for this approach to restoration. He called it intelligent tinkering. When a method derived from tinkering was successful, it was usually repeated, sometimes by others who were in communication with the tinkering practitioner. Eventually, the method became generally accepted and commonly used, perhaps with refinements and probably with modifications to accommodate different species and for use in different kinds of ecosystems. By this time, the method became common knowledge. Whether or not anyone bothered to describe the method formally in a published paper was immaterial. The method would eventually be described in the practitioner literature, usually without attribution to the tinker who first tried using it. This process has led to what is known as *knowledge-based restoration*. Nearly all ecological restoration projects depend largely or entirely on knowledge-based methods and strategies.

Practitioner literature is modest and not at all comprehensive. Papers are often sketchy, and well-crafted case histories wherein strategies are described and techniques and results are evaluated rigorously, are all too rare. In spite of these shortcomings, this literature has facilitated communication among practitioners. A significant boost to this literature was the launching of the journal, *Restoration & Management Notes*, founded and edited by Bill Jordan in 1983 and published by the University of Wisconsin Press. This journal gave practitioners an outlet for publication, and Jordan's poignant writings stimulated a growing restoration movement. The Society for Ecological Restoration (SER) held its first conference in 1989 in Oakland, California, and thereafter has provided a much-needed forum that drew practitioners together and greatly improved communications. There are still no instruction books or manuals to tell you how to design a project and what to do first. *The Tallgrass Restoration Handbook* (Packard and Mutel 1977) comes close, and so does *A Guide for Desert and Dryland Restoration* (Bainbridge 2007) with respect to prairie and desert ecosystems.

## Traditional Ecological Knowledge

There has been much interest in incorporating *traditional ecological knowledge* (TEK) into the restoration practitioner's toolbox. The assumption is that tribal people and others who live traditional rural lifestyles have learned how to manage their lands in order to subsist. Accordingly, restorationists can tap into this knowledge and improve their own skills. However, TEK arose from managing semicultural landscapes rather than restoring ecosystems. Therefore, TEK is more relevant to ecosystem management than it is to the recovery of ecosystems that were impaired by the environmental misdeeds of modern industrial societies. Traditional cultures cannot afford to impair their lands if they are to survive. Diamond (2005) painfully described many instances when traditional cultures collapsed on account of their misuse of natural resources.

The tactics and methods developed through intelligent tinkering treat the land much as would a traditional culture, the main difference being that modern implements are available to restorationists. Much knowledge-based information has been passed along by word of mouth from one practitioner to another, just as TEK is kept alive by an oral tradition. That knowledge may be spread at technical meetings rather than around tribal campfires.

Locating and making contact with people in traditional cultures is sometimes difficult, although SER's Indigenous Peoples' Restoration Network has made serious efforts to overcome that barrier and bring tribal and ethnic peoples from around the world to participate in SER's conferences and continue communications over the Internet. So far, TEK has yet to contribute significantly to our knowledge concerning project design and implementation, although it has been inspirational for restoration practitioners who want to recover ecosystems in ways that do not rely on technical solutions. We recommend that reference models include whatever local TEK is available that has relevance to project design and the development of project plans. Those responsible to preparing a reference model should seek sources of TEK. Our rationale is that TEK may open up options for conducting restoration that would not otherwise be obvious.

### Science-based Restoration

Gradually, scientifically inclined practitioners began designing and implementing ecological restoration projects using experimental designs and strategies that were amenable to systematic assessment. Knowledge-based methods, derived from intelligent tinkering that worked well in the past, were verified with empirical data. Their scientific validation was therefore no surprise. Nonetheless, such work gave birth to what is now known as *science-based restoration*. The high value that modern culture places on science makes us forget that nearly all ecological restoration resides on knowledge-based tactics and methods that were derived from intelligent tinkering and subsequently verified by science. Many papers published in SER's journal, *Restoration Ecology*, during the early to mid-1990s described restoration projects that were authored by scientifically minded practitioners and by scientists who were exploring ecological restoration as a potential new discipline. Papers of this genre still appear in the journal. Concomitantly, intelligent tinkering continues to be alive and well in restoration practice, where innovation is frequently needed to overcome technical constraints and pragmatic obstacles.

*Restoration ecology* as a significant discipline can be said to have come of age at a joint meetings of SER and the Ecological Society of America held in Tucson, Arizona, in 2002. Thereafter, many investigations have been conducted and published in *Restoration Ecology* and many other scientific journals of first rank, including *Science, Nature, BioScience, PLoS, Proceedings of the National Academy of Sciences, Ecological Applications, Ecology Letters, Journal of Ecology*, and *Frontiers in Ecology and the Environment*.

Much information in these publications is directed more toward exploring ecological theory than solving specific issues that practitioners face at project sites. Part of the reason is that funding agencies are only secondarily interested in practical applications. These agencies prefer to underwrite replicated experimental studies in which parameters are controlled much more carefully than can be expected in most ecological restoration project sites. Editors of prestigious scientific journals reinforce this concern. As ecological restoration becomes better known and more broadly accepted, however, this situation will change, and there are already early signs of that happening. For example, in mid-2004, the journal *Science* commissioned and published an entire special section of articles devoted to restoration ecology. A quarter century earlier, in the centennial issue of *Science*, published on July 4th, 1980, no articles were included on conservation, ecology, or restoration. This is a remarkable indication of the important steps toward mainstreaming ecological restoration and its sister disciplines. In spite of these advances, interest lags far behind in regard to issues pertaining to the concomitant socioeconomic and cultural aspects of ecological restoration. Only ~ 3 percent of all reports on ecological restoration projects published in thirteen peer-reviewed journals between January 2000 and October 2008 addressed the socioeconomic benefits—or lack thereof—of the projects (Aronson, Blignaut, Milton et al. 2010).

Science-based restoration is still very new and is only just beginning to realize the great promise it has to advance the practice of ecological restoration. There has been a recent explosion in university courses pertaining to ecological restoration and particularly restoration ecology. Some academic institutions offer certificates and degrees in ecological restoration. Much of this activity has been stimulated by government agencies, NGOs, consulting firms, and even private corporations that are employing practitioners. These trends are highly encouraging.

# Chapter 9

# Project Planning and Evaluation

The methods and tactics for restoring an impaired ecosystem are disarmingly simple. Some restoration projects may use earth-moving equipment that is automated by global positioning systems (GPS) and equipped with lazar-leveling devices, which may seem complex to some, but such technology is in common use in civil engineering. Site preparation may be conducted using subsoil rippers, seed drills, cultipackers, and automated tree-planting machinery, but such equipment is merely borrowed from standard practice in agronomy and forestry. For the most part, restoration is performed with relatively low-tech methods, which seems appropriate for a discipline that returns ecosystems to a former condition.

Nonetheless, ecological restoration projects are surprisingly complex and require considerable attention to detail. Like a jigsaw puzzle that is missing a few pieces, an ineptly designed restoration project can look simply awful, ecologically speaking, and not work at all when it is said to be completed. The analogy of a jigsaw puzzle is appropriate for ecological restoration, whereby all of the pieces, or steps along the way, are simple. However, they are interconnected and must fit together in a holistic manner. The job of the restoration planner is to identify all of the pieces and explain how they are to be put together ecologically to benefit biodiversity and fulfill values to the satisfaction of stakeholders and the local community.

A satisfactory ecological restoration project is one that was conceived on ecological principles and planned to translate those principles into appropriate action. Competent project implementation cannot be discounted; however, it must be predicated on proper conception and planning. This does not mean that a well-conceived and planned project will be carried to completion without incident. It does mean that practitioners should have the flexibility and support to overcome unanticipated hindrances as they arise and that such problems will be kept to a minimum. Much of the critical work, therefore, occurs as a new project is being conceived, prior to the time when the decision is made to initiate its actual implementation.

*Intervention* and *manipulation* are two terms that are commonly used interchangeably in discussions of planning, but we recognize a subtle distinction. A manipulation is a direct activity to improve biophysical attributes at a project site, such as amending the soil with lime or outplanting nursery-grown stock. An intervention can be an alternate term for manipulation, or it can designate another action that facilitates biophysical improvement indirectly. Excluding grazing by livestock is an indirect intervention that removes grazing pressure on desirable plant species and promotes their proliferation by natural regeneration, as was described in chapter 8 for the restoration of tropical dry forest. The intentional introduction of livestock on a restoration project site to reduce unwanted competitive grasses, as was described in VFT 2 in France, is a manipulation and therefore a direct intervention. The distinction is subtle but useful in describing restoration strategies.

## Guidelines for Restoration

The steps and tasks that comprise an ecological restoration project were identified and summarized in *Guidelines for Developing and Managing Ecological Restoration Projects* (Clewell et al. 2005). This document was adopted as a foundation document by the Society for Ecological Restoration and is known as the SER *Guidelines*, which is posted on the SER web site (www.ser.org). These fifty-one guidelines encompass project work from initial feasibility studies to the preparation of the final report for a completed project. The SER *Guidelines* are not the only source of guidance; Wyant et al. (1995), Whisenant (1999), and Hobbs (2002), among others, have provided useful contributions to help those preparing protocols to aid in decision making.

The SER *Guidelines* serve as a checklist for practitioners and project managers to ensure that they make no errors of omission. The guidelines also serve as a filing system for project information. A digital file can be made for each guideline. Project activities can be noted in the appropriate file as if it were a diary. Descriptions, data, photographs, and maps can be entered as they become available, as can other documentation and pertinent bibliographic references. These files document all aspects of the project in a format that facilitates the preparation of progress reports, midterm funding proposals, press releases, final reports, professional presentations, and publications. If new personnel join the project, the documentation allows rapid familiarization with the project and its current status.

The SER *Guidelines* are categorized into six groups: conceptual planning, preliminary tasks, implementation planning, implementation tasks, postimplementation tasks, and evaluation and publicity. We review them as follows.

## Conceptual Planning

Tasks in this group collectively represent a feasibility study to determine whether restoration should be conducted. The steps in conceptual planning begin with the identification of the project site on maps and aerial photos on which the project boundaries are drawn. Enough of the surrounding landscape is included to show the catchment, land use, and other features that could influence the project. The sponsors of the project site are identified along with their contact information. The kind of ecosystem that is to be restored is succinctly identified, along with the causes and degree of impairment. The justification for embarking on ecological restoration is given. The impaired ecosystem is envisioned the way it will appear after it has been made whole again and regains its functionality.

One of the most important tasks in conceptual planning is to identify stakeholders and interested leaders in the local community, in order to seek their endorsement and participation in the project and to engage them in the process of developing project goals. Project goals are succinct statements that describe the vision of the project. Ideally, the goals include recognition of the eleven attributes of a satisfactorily restored ecosystem, as described in chapter 5. Project-specific goals with regard to other values (chap. 2) may be added.

The next step is to identify the general tactics, but not the specific methods, to accomplish ecological restoration. Ecological drivers or stressors should be identified that will have to be reengaged or accommodated during the course of project work, for example, a prescribed burning regime. If for any reason these ecological drivers are not expected to continue indefinitely once active restoration activities at the project site have been completed, the project should be abandoned or its vision changed. There is no reason to undertake restoration if the restored ecosystem is not self-sustaining or if there is little probability for long-term management, such as conducting prescribed burning.

Concurrently, any constraints in the surrounding landscape should be identified that could hinder restoration activities or that will pose a threat to sustainability thereafter. For example, will the restored ecosystem persist and function if impending development is constructed on the adjoining property? Will there be sufficient flows and exchanges of materials and organisms to sustain the restored ecosystem in the face of such development? Will the surrounding landscape become a perpetual source of invasive species that could colonize the restored ecosystem? Will the fire regime be affected? If serious issues regarding the landscape context cannot be resolved, perhaps a decision to move ahead with the restoration project should be reconsidered.

Sources of funding, labor, equipment, and biotic resources need to be identified and deemed adequate for project completion. Biotic resources may include seed sources and nurseries that can provide planting stocks. Permitting requirements need to be identified and the permitting process understood sufficiently to realize the probability of obtaining required permits. Legal constraints must be identified, such as zoning restrictions, restrictive covenants, liens, and restrictions concerning ingress and egress. The duration of restoration work at the project site should be estimated to ensure that it can be completed within time constraints that may apply.

### Preliminary Tasks

Once conceptual planning has determined that the proposed restoration project is feasible, a number of preliminary tasks may have to be accomplished before the final decision is made to prepare project plans and conduct restoration. Some tasks are administrative, such as preparation of a budget in support of preliminary tasks, and others are technical. One of the first technical tasks will be to conduct an ecological inventory of the project site that describes the hydrology, soils, and other aspects of the physical environment and that documents the biota in terms of species composition, structure, and species abundance (Aronson et al. 1993a, 1993b). This inventory will be used in determining the extent of restoration and deciding on restoration methods. Later, the effectiveness of the restoration effort will be evaluated in terms of the degree of recovery from this initial, impaired condition. This inventory does not have to be exhaustive, but it should contain at least some empirical information on physical conditions and the biota with which the restored ecosystem can be compared. In addition, baseline monitoring may have to be initiated on critical parameters, such as seasonally variable depths to a water table or dissolved oxygen content of surface waters. Monitoring for these parameters would likely continue for the life of the project.

Of critical importance is photo documentation. Many photos should be taken, including some at permanent photo points that can be readily relocated, where photos with the same views can be taken annually throughout the duration of the project and perhaps beyond. Among the saddest moments in the careers of practitioners is when they realize—five or ten years into a project—they cannot find good photographs of their project site before and shortly after the work began. They will wish they had taken dozens of high resolution pictures with proper lighting—both close up and wide angle and at different seasons—and that the photos were dated and identified by location.

The history of impairment should be documented in terms of what caused it and when it occurred. Historical aerial photographs of different ages can be useful in determining or confirming the extent and timing of impacts. A reference model needs to be prepared, as described in chapter 6. The reference model is carefully written and contains all information (or references to such information) that will be needed to prepare detailed plans for project implementation. Autecological information should be gathered concerning the life history, *phenology*, and habitat factors for critical species, if such information is not already available. That information will be particularly useful in those parts of the world where *local ecological knowledge* is limited. In addition, plants of some species can be grown in experimental plots to determine effective methods of propagation and establishment.

At this point, enough information has probably accumulated to reassess whether or not previously established project goals are realistic. If not, the goals should be adjusted accordingly. It makes no sense to cling to goals that are not likely to be achieved or that do not justify the effort needed to underwrite and undertake an ecological restoration project.

For larger projects, especially those that require proof of successful completion, short-term objectives are devised, which can be attained by the time active work at the project site has been complete, usually within five or ten years. Preferably, these objectives should pertain to the first four attributes of a successful restoration project as identified in table 5.1. Such objectives would concern both the physical environment (e.g., attainment of hydrological stage-duration parameters or the accumulation of organic content in the soil) and to biodiversity (e.g., establishment of particular species in terms of their sizes and abundance). These objectives (often called *performance standards* or success criteria), and protocols that would demonstrate their attainment from empirical monitoring data, are generally required as provisions in permits and contracts. Successful attainment of these objectives infers that longer-term project goals will eventually be met as the restored ecosystem continues its unassisted development (Wyant et al. 1995). The attainment of objectives is evidence that permit and contract criteria have been satisfied. Monitoring and protocols for data analysis should be established in advance of project work in order to provide unassailable evidence of the attainment of each objective. Permit applications commonly require detailed project plans to be attached. There may be no need to establish objectives for small projects and for those conducted as components in environmental stewardship programs that have permanent staff members or volunteers who can provide extended aftercare, if it should be needed.

By this time, liaison should have been established with all interested parties, including stakeholders, the local community through its leadership, and news media; also public agencies, nongovernmental organizations with interests in natural resource management, conservation and environmental groups, and educational institutions. These parties can be helpful by endorsing the project and by providing support in terms of funding, labor, expertise, and equipment. Additional preliminary tasks may entail the preparation of access roads, electric lines, and other infrastructure in support of onsite restoration project work.

## Implementation Planning

At this point, sufficient information has accumulated so that plans for the conduct of restoration tasks can be prepared. These are usually called project plans; however, we call them implementation plans to distinguish them from conceptual and preliminary planning tasks. The restoration planner can now be engaged. The planning process identifies and describes each restoration task that will be needed to recover the biophysical environment from impairment. Among these tasks will be those that fulfill the short-term objectives. The project planner prepares drawings and descriptions of tasks to show what work is to be accomplished and how it will be implemented. Implementation tasks in restoration project cover a wide array of issues, and they may differ widely among projects. Figures 9.1 and 9.2 illustrate one such issue pertaining to reducing the environmental impact of an access road that traverses a restoration area. This is the kind of task that is not often considered, but it can change a routine project into one that is sophisticated ecologically.

Once plans are completed, a budget for conducting restoration is prepared, including funds held in reserve for unanticipated contingencies. By this time, the project manager has been engaged who will secure equipment, supplies, biotic resources, and labor. Restoration personnel may require training, particularly volunteer labor. Perhaps the most critical role of the project manager will be to schedule project activities in proper sequence in accord with growing seasons and the phenological requirements of key species. Labor and equipment must be scheduled for availability in accord with this ecological calendar. Suppliers of biotic stocks must be carefully instructed, so that stocks have reached their peak conditions of development when delivered. In spite of such careful planning, delays may be inevitable on account of inclement weather, equipment malfunction, administrative snafus, regulatory delays, and other aggravations that should be anticipated and avoided insofar as possible by the project manager.

## Implementation Tasks

Implementation of project plans onsite is performed or overseen by restoration practitioners. There may be additional work, such as marking boundaries and installing fencing. Stakes or monuments may be installed to mark permanent

**FIGURE 9.1.** Typical graded forest road drained by a ditch on either side in Florida, USA. The ditch and slightly elevated road intercept surface runoff of precipitation and modify the hydrology in adjacent pine savanna.

monitoring locations and photo documentation points. In some projects, implementation is completed in one event that may span a growing season or a similarly finite period. In other projects, implementation may be pulsed as two or more discrete events that could be separated by several years, for example, to allow the development of shady habitat prior to introducing species that are intolerant of open environments.

### Postimplementation Tasks

The newly implemented restoration must be protected against vandals, herbivores, and other threats until growing stocks have become established. Mice, geese, deer, feral pigs, nutria, and many other animals view a newly planted restoration site as dinner. Teenagers view the same site as a race track for all-terrain vehicles.

The most important postimplementation task is to provide aftercare, which consists of those biophysical interventions by practitioners to nurture the recovering ecosystem until it can cope independently. Some aftercare tasks may be prescribed in project plans, such as the irrigation of outplanted nursery stock.

**FIGURE 9.2.** Forest road near the one shown in fig. 9.1 in which the side ditches were filled and the roadbed graded to restore hydrological processes.

The need for additional aftercare may become apparent from reconnaissance following implementation to identify unanticipated problems that necessitate midcourse corrections. *Aftercare* is sometimes called maintenance or management. We prefer the term aftercare to designate postimplementation actions taken to ensure that all essential ecological processes have returned to normal levels of functionality and that assistance from restoration practice is no longer needed. We apply the term management (in the sense of ecosystem management) to activities scheduled after all restoration tasks are finished and the restoration project has been completed.

Aftercare can take the form of adaptive management (as opposed to ecosystem management), whereby the biotic responses to implementation will determine what kinds of aftercare will be attempted. The prescription for aftercare is tailored to the response of the biota to initial implementation. Thereafter, the next prescription for aftercare will be based on the response to the previous prescription. Each aftercare event is conducted as if it were an experiment, much as a physician would prescribe pharmaceuticals serially until the patient responded positively to one of them. Project plans can authorize adaptive management activities in advance and allow adaptive management to be budgeted. This procedure may be more palatable from an accounting perspective than to include a line item in a budget for undisclosed contingencies.

Scheduled monitoring is conducted as a postimplementation task. Monitoring is needed to determine if and when short-term objectives have been attained. If monitoring data indicate that attainment is not forthcoming, aftercare in the form of midcourse corrections may be warranted. Monitoring that is not required to measure the attainment of specific objectives is costly and may be difficult to justify in terms of funding. However, without monitoring, the success of a restoration project may be difficult to document. For that reason, the selection of short-term objectives and monitoring protocols to determine their attainment during the preliminary project planning stage take on a greater importance than merely determining whether or not a subcontractor should be paid for services rendered.

### Evaluation and Publicity

As a restoration project draws to a close, monitoring data are evaluated to determine if and when performance standards were attained. A final report is prepared that describes the project as a case history. Technical presentations are made and articles prepared for publication so that other restoration professionals can learn from the achievements—and the mistakes—of the project. The news media are informed to announce project completion, and a public celebration can reinforce publicity and favorable public attitudes toward the completed project. A celebration could be a community-wide picnic with tours of the project site, congratulatory speeches by dignitaries, and live music. The importance of publicity, permanent signage, and public celebration cannot be overemphasized. Otherwise, the local community may not realize that the project had come to fruition and that the benefits accruing from the restored ecosystem were attributable to a well-conceived and executed restoration project.

## Strategies and Designs

Restorationists commonly speak of designing a project and of a project's design. Design is a term that is appropriate for engineers and architects whose end products are predictable and exact. It is not as readily applicable a term when applied to ecological restoration, in which practitioners have only a few years to reengage ecosystem processes to the point that self-organization begins. Instead of engineering and architecture, ecological restoration is more closely related to pediatrics. The concerns of both fields are youths—young people and youthful ecosystems—that are just beginning their development in a manner that defies the precise prediction of final outcomes. Practitioners in both fields have targets at which they aim in terms of norms of physical development and role models of behavior for youths and reference models for impaired ecosystems. However, not all youths meet physical and behavioral norms when they become adults, and not all ecosystems meet their intended targets, for reasons that were described earlier, primarily in chapter 4. A design promises or at least suggests that a particular endpoint will be achieved. Ecological restoration is open ended and cannot make that promise with a high degree of certainty, except for ecosystems that need little intervention and will recover from impairment very rapidly.

Project design therefore is a somewhat misleading term in the context of ecological restoration, despite its convenience and its widespread use. We prefer to speak of a restoration project in terms of its strategy (overall approach to accomplishing restoration) and tactics (methodological approaches selected to carry out a strategy). Here the inference is guidance rather than insistence. We are helping a young ecosystem to reestablish its historic continuity. We really have no control over what will happen to it in the long term, just as we cannot control our children's lives after they strike off on their own as young adults. We can inculcate our children with moral behavior and a good education, just as we can imbue degraded ecosystems with favorable hydrology and desirable species. But we cannot say for sure how either will ultimately turn out. Engineers and conventional architects are not concerned with living, open-ended systems. Their only interest is that the bridge does not collapse and that the building is efficient and aesthetically pleasing. In other words, only end products count. Many restoration practitioners have been inspired by Ian McHarg's (1967) concept of design with nature. This phrase has become the byword of an entire generation of landscape architects cum restoration practitioners (Orr 2002; Todd 2005; Apostol and Sinclair 2006; see also the writings of Richard Forman, e.g., Forman 1995). Nonetheless, we suggest that strategy is a more appropriate term. From the standpoint of accuracy and effective communication, we hope that it will become more widely adopted.

## Genetic Provenance

An important aspect of restoration planning is the genetic provenance—or place of origin—of seeds, other nursery stock, or other organisms that are introduced at a project site (Falk 2006; Maschinski and Wright 2006). The reason is that many species are wide ranging and occur in regions with a different climate, soils, or other environmental variables. Consequently, organisms from another locality that are introduced in a restoration site may have low survival, poor growth, low vigor, low reproductive capacity, or low tolerance to extremes in weather. Most individuals of a species are well adapted to the place where they live because of pressures of natural selection for particular phenotypes and the alleles that produce them, which favor high survival and successful reproduction under local environmental conditions. Adaptation is expressed in terms of physiological response, such as cold hardiness, drought resistance, or the timing of phenological events such as the breaking of dormancy. These differences are rarely linked to phenotypic expressions that are detectable from morphological examination. For example, the presence of a floral bud does not indicate when it will open. However, reciprocal garden studies allow the detection of an ecotype, consisting of species populations that are adapted to a specific set of environmental conditions and that usually occur in a specific location or genetic provenance.

Next to nothing is known about ecotypic differentiation in all but a few species. Little is known about the gene frequencies of adaptive alleles within a given ecotype and the number of generations needed for an ecotype to differentiate; however, the principle of ecotypic differentiation is widely accepted. Restoration ecologists universally assume the existence of numerous ecotypes and regulatory agency personnel commonly insist that planting stocks be obtained within an arbitrarily set distance from a restoration project site. Introductions of unfavorable ecotypes can cause biological havoc, such as flowers being produced in the wrong season for local pollinators or plants breaking dormancy during the dry season. Furthermore, hybridization may occur among introduced ecotypes, leading to genetic erosion of the local populations.

If the intent of a project is to emulate the preimpairment state, then planting stocks should be obtained locally to ensure that all stocks represent local ecotypes. If the intent is to restore an ecosystem on mined and physically reclaimed land, where the soil structure, hydrology, and perhaps other environmental conditions were altered relative to the premining environment, then it would be best to introduce multiple ecotypes. Selection pressures for the new, postmining environment would determine which ecotypes survived and could also initiate natural selection for a new ecotype from the array of alleles in the genotypes that were present. If the climate is predicted to become warmer and drier, then planting stocks should be obtained from another region where the climate is already warmer and drier. This option is not intended as a form of assisted migration but rather as a way of preadapting a restored ecosystem for anticipated environmental change.

If local ecotypes would improve the opportunities for successful restoration, much care should be exercised with regard to the genetic provenance—or location—of where planting stocks are obtained (Krauss and Hua He 2006). For example, in mountainous regions, stocks should come from nearly the same elevation and aspect, and from soils that originated from the same kind of parent material. A difference of a few meters in distance could be critical in locations where localized soils were derived from serpentine outcrops. However, in a large region with uniform environmental conditions, ecotypic differentiation is much less likely to be pronounced, and a local ecotype may be assumed to occupy thousands of square kilometers.

Practitioners should be cautious when dealing with, or buying stock from, growers in commercial plant nurseries to ensure that a suitable and fully documented genetic provenance of plants is being employed. Growers commonly buy seeds from wholesale seed collectors, who obtain them from distant locations and from any and all suppliers. As a general rule, planting stocks should not come from regions with a wetter, drier, or warmer climate than the one at the restoration site. Practitioners should also avoid purchasing seeds or seedlings of native trees that have been genetically selected by plant breeders for timber production. Such trees may have had alleles bred out of them that are important for certain ecological functions, such as the capacity to form hollow trunks or other deformities that animals need for nesting and denning.

## Inoculating Soils and Substrates

Well-developed soils generally contain a wealth of species—fungi, bacteria, algae, insects, other arthropods, protozoa, nematodes, annelids, and so forth—most of which are not inventoried by restorationists for lack of taxonomic expertise, laboratory facilities, time, and financial resources. Yet these organisms are important ecologically for their roles in decomposing organic matter, recycling nutrients, maintaining soil texture, aeration, and other crucial processes. In sites where soils were damaged or recently reclaimed, these species can be readily reintroduced by salvaging soils from donor sites scheduled for conversion or *reallocation*. Salvaged soil is spread on restoration project sites. This technique has been called "topsoiling" or "mulching" and if it is practiced on relatively flat terrain, it is commonly followed by disking or other mechanical treatments to help incorporate the transferred soil, soil-borne seeds and rootstocks, arthropods, and microorganisms into the substrate at the project site. On steeper slopes with erosion potentials, topsoiling may have to be limited to safe sites (chap. 4). If the supply of donor soil is limited, or the cost of collection and transport excessive, small quantities can be transferred to a project site and incorporated into the substrate in strategic locations, where desirable biota in the transferred soil can radiate into surrounding soil. Similarly, translocated sediments from an aquatic donor site can inoculate benthic macroinvertebrates into a stream or wetland that is undergoing restoration.

The project site must be suitably prepared to receive donor materials. There is no benefit to be gained from transferring donor soils from a moist habitat to a desiccated restoration site where its organic matter content will oxidize (Clewell et al. 2000). Where demand for restoration exceeds supply, an explicit goal of a restoration project can be included to establish a source of donor soil for use as inoculum at future project sites.

## Project Evaluation

Critics of ecological restoration say, all too often with justification, that too little monitoring is required or it is never conducted. We concur that monitoring generally needs to be upgraded and perhaps combined with postproject ecological evaluations in order to demonstrate project completion, provide accountability to financiers, and provide information that informs technical personnel charged with prescribing future ecosystem management at the project site. In addition, monitoring and evaluations can inform future ecological restoration practice on the effectiveness of restoration methods and provide information to those who will use the site for purposes of education or developing social capital. Finally, all empirical data are potentially useful for research purposes by restoration ecologists.

Providing proof that a project was completed may seem frivolous. However, a successfully completed project is obvious only to those who were familiar with the impaired project site before restoration began. Others who are shown the site may not realize that natural area beneath their feet had been restored. The most effective way to demonstrate that a project was completed is with before and after photo documentation, preferably as shown in a report that evaluates the project.

Demonstrating accountability is important to assure financiers of project completion. Potential sponsors, financiers, and policy makers need to know that restoration projects can be conducted successfully (Bernhardt et al. 2005). They need to have a realistic assessment of the costs for conceiving and planning future projects. For that reason, project costs should be summarized in the evaluation process, as they were in VFTs 1 and 7. The basis for determining project costs must also be specified. For example, the person hours of volunteer labor (if any) should be noted or estimated along with calculations of restoration costs per acre or hectare. Otherwise, cost estimates can be misleading and grossly misinterpreted.

Project evaluations are crucial for the development of the profession of ecological restoration. Practitioners need to know what strategies and methods have worked well and under what conditions. They also need to know what has not worked well and why, so that mistakes are not repeated. Practitioners need to share their experiences with their peers by providing evaluations of their projects in readily accessible places, such as websites and journals. The preparation of a case history of the project is vital in this regard. A case history should be prepared in a manner that allows comparison with the reference model and facilitates assessment of how well attributes of whole ecosystems (table 5.1) were attained.

Elements that contribute to a case history are an ecological description of the impaired ecosystem prior to restoration, the reference model or a summary of it, any baseline data, a chronological record of interventions by practitioners, project monitoring data, photo documentation, and a summary of project costs.

The conclusions drawn in any evaluation of an ecological restoration project must be prepared carefully, because there are different norms by which a restored ecosystem can be evaluated ecologically, which can lead to different conclusions. Evaluation must be based on project goals rather than unrelated criteria. For example, Craft et al. (2002) evaluated tidal marsh restoration and determined that marsh grasses approached the stature and abundance of those in an undisturbed reference marsh in only a few years but that soil organic matter would take several more decades to be restored. Should the restoration be hailed as a success based on vegetation or deemed a failure on account of soils? The answer to that question depends on the goals of the restoration project. If the project goal was to provide primary production for estuarine food chains, the project was an immediate success because of the abundance of marsh grasses. If the goal of the project was to sequester carbon from the atmosphere to reduce global warming, the project is not yet successful because of the low content of organic matter in the soil.

In another example, Zedler and Langis (1991) reported that tidal marsh restoration was inadequate because the dominant grass (*Spartina foliosa*) had not attained sufficient height to support nesting by clapper rails. Later, Boyer and Zedler (1999) explained that the reason for failure was that the site was constructed at an inappropriate elevation for *S. foliosa* to attain its maximum height. Was anyone at fault? Was it the engineers who graded the site to the wrong elevation before restoration activities began? Or was it the permitting agency for specifying the wrong elevation? Or was the restoration practitioner to blame for planting inappropriate species for that elevation? Or was the restoration really successful but site conditions were inappropriate for clapper rails, and they should not have been used as a criterion for success? Were other factors involved that were overlooked by the evaluation team? Was nobody at fault, because our knowledge at that time had not yet been calibrated by experience? In this instance, different investigators using the same data could come to diametrically opposed conclusions regarding the worth of ecological restoration.

The use of reference models in evaluations requires judgment. We have already emphasized that ecological restoration is openended and that the reference model serves as the starting point for a project but not necessarily the endpoint. Almost invariably a reference model represents a mature stage of ecological development, whereas a recently completed restoration project is much less mature. This complicates comparisons.

If the evaluation is delayed until the restored ecosystem has reached the same degree of ecological maturity as its reference model, then the comparison could be compromised by circumstances that affect ecosystem development after the completion of restoration activities. We report here a particularly poignant example of this situation with regard to 8 hectares of restored forest on previously mined land along Dogleg Branch in Florida (fig. 9.3) (Clewell et al. 2000; Clewell and Aronson 2007, 141–46). The restoration plan was approved by the predecessor agency of Florida Department of Environmental Protection (FEDP), which certified that restoration had been successfully completed thirteen years later in 1996. The project site was deeded by the mining company to the State of Florida in 1996, and it became part of the Alafia River State Park, which is administered by FDEP. A prescribed fire by park personnel unintentionally burned 0.4 hectare of restored forest in 2004. Our reconnaissance in 2010 revealed recent establishment by alien invasive species and fresh damage by feral pigs that had consumed much herbaceous vegetation and some sapling trees. The restoration cannot be evaluated on the basis of its current condition, except with interpretation and professional judgment, because the site has been compromised by lack of protection and ecosystem management by the same public agency that approved the restora-

**FIGURE 9.3.** Forest along Dogleg Branch, Florida, 27 years after ecological restoration began.

tion plans and certified their satisfactory completion. In fairness to FDEP, state parks have been underfunded in recent years.

Studies could be directed at evaluating a restored ecosystem as natural capital (chap. 10) and its ability to provide ecosystem services. For example, is stormwater runoff retained on site and flooding reduced downstream? How many kilograms of fuelwood can be harvested per year by villagers? Do teachers bring students to the restored ecosystem for nature study? Does the restored ecosystem support populations of rare species that were specified for introduction in project goals? These are the assessment criteria that are meaningful to stakeholders and ultimately to policy makers and financiers who will decide whether future restoration projects will be authorized and underwritten. They deserve rigorous documentation. The degree to which a restored ecosystem resembles its reference cannot be discounted, but it is ultimately of secondary importance relative to the value of that ecosystem to people.

Dean Apostol and Jordan Secter

The northwest coastal region of the United States is arguably among the most natural places on Earth. Beneath the surface all is not so well in what has been called "ecotopia." In particular, the beautiful rushing rivers with abundant cold, clear water mask a steep decline in native fish, particularly salmon. The Siuslaw River is one of them. Causes of salmon decline include overfishing, construction of dams, loss of stream habitat, and competition with hatchery-raised fish.

The 200,000 hectare Siuslaw River watershed drains a lightly populated area in the central Coast Mountain Range of Oregon to the Siuslaw estuary near Florence. This watershed is entirely free of major dams and has few natural barriers to salmon migration. Historically it had the second highest production of coho salmon in the state, with over 200,000 adults estimated in the late nineteenth century. The population dwindled to only a few thousand adults by the mid-1990s (Ecotrust 2000). Deterioration of stream habitat caused the decline of coho. The Siuslaw River watershed was heavily impacted by logging, farming on the narrow valley floors, a legacy of temporary "splash dams" that flushed gravel and wood from channels, and an increase in debris flows linked to road building across unstable slopes. Many streams were down cut, some all the way to bedrock. Log jams that held sediment and nutrients in place were destroyed and cannot reform because of a lack of large woody debris (Ecotrust 2000).

These impacts simplified the aquatic ecosystem. Streams in downcut channels lost contact with floodplains and wetlands. Large wood, a keystone element of northwest aquatic ecosystems, is mostly absent, and the few pieces of wood that fall into streams are quickly swept to the estuary and out to sea during winter storms. The lack of habitat complexity expedites nutrient leakage. Young salmon consume aquatic invertebrates in a food web that depends on leaf litter as the carbon source. If leaf litter is not retained by log jams, wetlands, and back channels, salmon populations collapse.

In addition, the estuary was damaged. Nearly 60 percent of its wetlands were lost to channel dredging. A navigation jetty caused wood and nutrients to be funneled rapidly into the open ocean. Salmon were unable to increase their body fat in the estuary before heading out to sea, which reduced their chances at ocean survival.

Efforts at salmon enhancement, which began in the late 1960s, were often counterproductive, particularly removal of beaver dams and log jams in the mistaken belief that these blocked salmon migration. Knowledge and methods improved, as evidenced by a restoration project at Karnowsky Creek, which joins the Siuslaw River Estuary about nine miles east of the Pacific Ocean. Siuslaw Indians lived along this creek and throughout the watershed for thousands of years and harvested salmon sustainably with weirs that allowed the passage of some salmon to upstream kin and spawning areas. Euro-Americans settled in the Karnowsky Creek area in the late 1800s. They cleared old growth coniferous forest from the valley floor, drained wetlands, and built dikes to hold back tidal waters. They farmed and tended livestock. Karnowsky Creek, which once meandered down the middle of the valley, was redirected to drainage ditches dug into valley edges, and the original channel filled with sediments.

Channelizing destroyed salmon habitat by reducing ecosystem complexity, removing contact with the floodplain, and lowering the water table. Summer water temperatures became too high for salmon, and they lost off-channel refuges that provided escape from summer heat and winter high flows. Tide gates blocked migration.

Like many small farms in the Oregon Coast Range, those in Karnowsky Creek eventually became uneconomical. The US Forest Service purchased the valley bottom in 1992, with the intent of restoring the creek as salmon habitat. In 2001, the Forest Service and the citizen-based Siuslaw Watershed Council joined to encourage stream health. This coalition sponsored the restoration of the original meandering channel and its wetlands, as well as the resumption of normal tidal flows in Karnowsky Creek. They raised funds and recruited an interdisciplinary team of students from two nearby state universities to prepare the restoration design. The team was given two months to gather information, develop restoration strategies, and present their proposal to the community.

The design team prepared a detailed topographic map of the valley. They measured cross sections of the ditches and recorded storm flows at the mouth. Sections of the original stream channel were identified from historic and current aerial photos. Intact Hoffman Creek nearby served as a reference site to inform restoration design. The team measured Hoffman Creek's cross sections, pool dimensions, meander curvature, pool to riffle ratio, and gradient. Hydrologists calculated that the ditched version of Karnowsky Creek was sized correctly to carry normal flows. Designers purposely undersized the restored creek, so that it carried approximately two-thirds of that volume. This procedure would allow its channel to self-adjust to the flow regime over time.

The final report was widely accepted in the local community. The Forest Service, Siuslaw Soil and Water Conservation District, and the Siuslaw Watershed Council received > US$400,000 in grants to restore > 3 miles of the original stream channel and adjacent wetlands, floodplains, and areas of tidal influence. Construction began in August, 2002, and was completed two years later. The meandering channel was excavated and old drainage ditches plugged (photo 1). Logs of large trees were transported by helicopter and placed on the floodplain and in the channel. Local volunteers, including students and teachers from local schools, planted riparian trees, shrubs, and wetland vegetation. Students grew some of these native plants and participated in monitoring by taking water samples and measuring groundwater depths in wells.

The restored creek is expected to extend chum salmon habitat into the lower half mile. Coho salmon will migrate farther upstream and are ex-

**PHOTO 1.** Channel construction.

**Photo 2.** Hydrologically restored valley.

**Photo 3.** Young forest developing along Karnowsky Creek after restoration tasks had been completed.

pected to use adjacent ponds and connected floodplain refugia in summer and winter. Steeper channel sections are being restored to help provide spawning gravel.

In 2004, the Siuslaw Basin Partnership received the prestigious Thiess International Riverprize, awarded annually by an Australian organization for excellence in collaborative river restoration. As one of the featured projects in this award, Karnowsky Creek is now an international example of how agencies, special interest groups, schools, nonprofit organizations, and local residents have melded to collaborate as land stewards.

The Forest Service announced that the riparian plantings had succeeded beyond expectation by 2008, with willows and alders now exceeding ten feet tall and above the "free to grow" stage (photo 2). Beaver have returned to the creek. The newly meandered creek has retained its channel shape and location. Two new meander cutoffs are forming, which adds welcome ecological complexity. Groundwater has risen, and remains high later into the summer dry season than it did previously (photo 3). Groundwater cools water in the creek, to temperatures favoring salmon. A seasonal flooding regime has returned. Coho smolts became abundant a few years after the channel was restored.

Christian Little, Antonio Lara, and Mauro González

The Reserva Costera Valdiviana (RCV) is located in the mountainous Valdivian Rainforest Ecoregion of southern coastal Chile (39°58' S, 73°35' W). It is recognized as a biodiversity hotspot and it was assigned to one of the highest conservation priority rankings worldwide (Olson and Dinerstein 1998). The RCV is privately owned by The Nature Conservancy (TNC) and was created in 2003 to protect forest ecosystems. It initially covered 60,000 hectares until part of it was transferred to create a national park.

Native vegetation at the RCV consists primarily of uneven-aged and multitiered, temperate, coastal, broadleaved evergreen rainforest. Among the characteristic tree species are *Nothofagus nitida* and *N. dombeyi* (Fagaceae); *Drimys winteri* (Winteraceae), *Laureliopsis philippiana* (Monimiaceae), *Eucryphia cordifolia* (Eucryphiaceae), *Saxegothaea conspicua* and *Podocarpus nubigena* (Podocarpaceae); *Aextoxicum punctatum* (Aextoxicaceae); and *Amomyrtus mli* and *A. luma* (Myrtaceae). Forests of *Fitzroya cuppressoides* (Alerce; Cupressaceae) also occur in the RCV, and they contain some individuals that exceed 3,000 years old. *Fitzroya* is a monotypic genus endemic to southern Chile and Argentina. The mild temperate climate of the RCV is characterized by annual rainfall of 2–4 meters, with a pronounced austral summer dry season between January and March, during which time only ~5 percent of the yearly precipitation falls.

Between 1999–97, > 3,000 hectares of second-growth forests (~ 150 years old) and some old-growth forests (> 400 years old) in the RCV were clear-cut, burned, and replaced with industrial plantations of the introduced Australian tree, *Eucalyptus globulus*. Today, these dense, monospecific, and even-aged *Eucalyptus* plantations are fifteen to eighteen years old and 10 to 20 meters in height (photo 1). Conversion to *Eucalyptus* plantations has been detrimental to flows of several ecosystem services, including provision of water (quantity and quality), nutrient retention, provision of habitat for biodiversity, aesthetics, and tourism opportunities (Lara et al. 2003, 2009; Little 2011; Nahuelhual et al. 2007).

A consortium was created in 2009 to conduct ecological restoration on the RCV for the recovery of native forests. The partners in this consortium consist of nongovernment organizations; The Nature Conservancy; the Universidad Austral de Chile in nearby Valdivia through its Faculty of For-

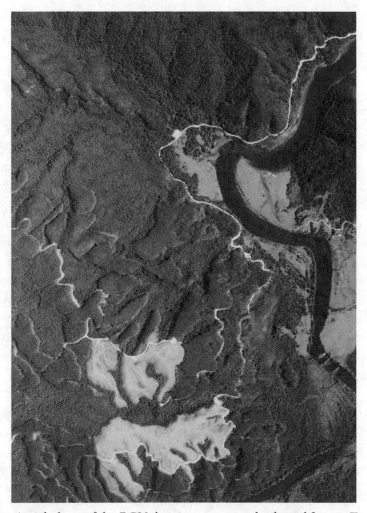

**PHOTO 1.** Aerial photo of the RCV showing two recently cleared former *Eucalyptus* plantations (light gray), several adjacent *Eucalyptus* plantations (dark gray, appearing raised), and intervening natural land in early stage of spontaneous forest regeneration.

est Science and Natural Resources; and MASISA S.A., which is a major commercial grower of exotic tree plantations in Latin America). MASISA is responsible for *Eucalyptus* removal at the RCV and is committed to responsible corporate standards regarding environmental and social issues. In particular, MASISA wants to increase its commitment to ecological restoration in order to increase its knowledge for the certification of its wood products by the Forest Stewardship Council. Lastly, FORECOS Foundation, dedicated to research, training, outreach, and networking of ecosystem services, is also an important partner in the project. The restoration project is being developed, integrated, and closely linked to a long-term research vision (Anderson et al. 2011). Financing for other aspects of the restoration program come principally from several agencies of the Chilean government; during the period 2006–12, funding was approximately US$400,000.

The community at large living in the small village of Chaihuin is gaining since one of its main activities is tourism. Increasing the landscape, biodiversity, and conservation value of the Reserve through restoration will promote tourism. Stakeholder organizations from local communities (i.e., the neighborhood council of Chaihuin and Rural Drinking Water Committee of Chaihuin are strong supporters of the project. Another important partner in the project is the Agricultural and Forestry Committee of Chaihuin (COAFOCH), a cooperative formed by local fishermen and farmers to ensure that social and cultural dimensions are incorporated into the project. Stakeholder support was instrumental in resolving dust issues and safety hazards in the community of San Carlos, which were related to the transportation of wood chips from the harvested plantations in the restoration project, by linking these issues to negotiations for ecological restoration at the RCV.

The recovery of native forest trees depends on an active ecological restoration plan because of competition from the much faster-growing *Eucalyptus*, their longevity, and their capacity for seed reproduction and coppicing. This landscape-scale restoration project is scheduled for completion in thirty years and will consist of the gradual reconversion of the *Eucalyptus* plantations to native forests that display the ecological attributes as recognized by Society for Ecological Restoration (SER 2004). The reference model for restoration planning was based on studies of the structure and composition of remnant uncut forests dispersed across the landscape. These second-growth forests are ~ 150 years old (photo 2). A conceptual ecosystem reference model included forest stand-development stages (Borman and Likens 1979; Oliver and Larson 1990), disturbance regime at a landscape scale, and social dimensions that determines the resilience in the long-term periods.

The reconversion of plantations to native forests started in 2010–12, with the removal of 50 hectares of *Eucalyptus* plantation. Another 100 hectares will be converted by 2014. Tree removal was accomplished by highly mechanized clear-cuts on 40 hectare tracts, and by less intensive cuts using

**Photo 2.** Temperate evergreen rainforest (~ 150 years old) and a reference site for ecological restoration.

**Photo 3.** Members of COAFOCH clearing *Eucalyptus* using oxen.

**Photo 4.** Restoration-project site showing a PVC pipe that marks a permanent vegetation-monitoring plot location and a recently outplanted sapling of *Nothofagus dombeyi* to the left of the pipe.

chainsaws and oxen on 10 hectare tracts by COAFOCH (photo 3). Nursery-grown seedlings of native trees were outplanted (1,500 seedlings/hectare) on cutover tracts. They consisted of *Nothofagus dombeyi*, and to a lesser extent *N. nitida* and *Drimys winteri* (photo 4). In the near future, seedlings will be produced using local provenance seeds of additional species. Seedlings are grown in a nursery that was established by stakeholders in a nearby community. Reforestation costs incurred for the first 45 hectares planted by MASISA are estimated at US$1,560/hectare and will be covered by income generated from the sale of harvested *Eucalyptus*.

A period of four to six years is anticipated to assist the reestablishment of native forests. This assistance includes the chemical control or removal of resprouting *Eucalyptus* stumps and any other potential invasive species, planting of native species to increase their density, encouragement of the natural recruitment of other successional forest species, and fencing to exclude cattle. The undergrowth, composed of tree, shrub, and fern species of

Photo 5. Natural regeneration of forest undergrowth and some native trees at a former harvested *Eucalyptus* plantation. Small tree with yellowish leaves in the center is *Drimys winteri*.

the original native forests, regenerates copiously from seed, and we anticipate no need to assist in its recovery (photo 5).

Monitoring has already begun and will continue for the next thirty years. In order to provide a baseline for water yields, preproject monitoring was conducted from 2006 to 2010 by measuring stream flow in nine watersheds using V-notch weirs (photo 6; Little and Lara 2010). These watersheds are covered variously by *Eucalyptus* plantations and native forests (Little 2011). Dissolved sediments, nitrogen, and phosphorus are monitored in the streams. These measurements are continuing to document the effect of restoration on water yields in a long-term monitoring program. Permanent plots installed before the *Eucalyptus* were cut are monitored biannually to follow the composition and cover of tree and undergrowth species, and to document survival of planted native trees (photo 4). We plan to expand monitoring to include bird and small mammal populations; macroinvertebrates in streams; and impacts of a socioeconomic, cultural and educational nature.

Photo 6. V–notch weir in a stream to measure water yield.

Hundreds of undergraduate and graduate students that have visited the restoration project in their field excursions, and as opportunities to do their seminars and theses. The project has been well publicized on radio, TV, and in newspapers, and it has already attracted numerous visits by government officials. The overall benefit of this project is expected to reach far beyond its boundaries and is anticipated to have a major impact on policy for the reconversion of pine and eucalyptus plantations across extensive areas in south-central Chile.

# PART IV

# Ecological Restoration as a Profession

Part 4 addresses the state of the emerging profession of ecological restoration. Chapter 10 examines the boundaries of ecological restoration and how it interfaces with similar disciplines and activities that attempt to achieve environmental improvements and to reduce environmental blunders and disasters. Chapter 11 identifies and describes project sponsors, stakeholders, and the various professionals who participate in restoration projects. Different modes of project administration are identified, and practitioner training and professional certification are discussed. Chapter 12 identifies three, sometimes conflicting and sometimes complementary models of the way professionals perceive ecological restoration. Several "hot-button" issues are explored in light of these models. The chapter and the book conclude with recommendations on how we may advance beyond the misunderstandings generated by the many paradoxes of our discipline and move forward in unison.

# Chapter 10

# Relationship of Restoration to Related Fields

Effective *ecological restoration* is one of a number of strategic initiatives that are urgently needed to counter and, if possible, reverse the accelerating deterioration of the biosphere and its wealth of natural resources on which all people, and all life on the planet, depend. There are many forms of actions—and associated professions—that have, at times, been conflated with ecological restoration. Insofar as they address environmental and ecological concerns, many of them are indeed related to restoration, but there is also a blurring and fuzziness at times. Cleaning up an oil spill, in and of itself, is not restoration; it is the elimination of toxic pollution. Planting turf grasses on a retired mine site or closed landfill may provide an attractive lawn or meadow, but it is not restoration. Proposed efforts to slow anthropogenic global warming through geoengineering of the atmosphere and oceans are also not ecological restoration but rather high-risk gambles we can all do without.

Policy makers, administrators, and others who need to know what each environmental discipline can offer are frequently misinformed about core concepts and the key words employed related to ecological restoration. Some of the confusion begins with the diversity of interests that restorationists themselves bring to our young and volatile discipline. Restoration is used so loosely by some authors and in some arenas that its meaning can be lost or misconstrued. Our goal is to raise ecological restoration from its current status as a buzzword so that it can achieve its full potential as a prominent element in national and international politics and policy.

We begin this chapter by exploring the differences between ecological restoration and *restoration ecology*. Next, we consider the relationships between ecological restoration and ecosystem management, rehabilitation, reclamation, *revegeta-*

*tion*, and remediation. Spreading our survey further, we consider compensatory *mitigation*, ecosystem fabrication and creation, landscape architecture and design, and ecological engineering. Some authors consider all of these disciplines and activities, including ecological restoration, as aspects or components of ecological engineering. Others, including ourselves, consider them as complementary and sometimes partially overlapping. We argue that we need to draw clear boundaries between disciplines in order to facilitate effective communication and collaboration as part of a concerted effort to resolve the truly daunting environmental problems that face us. This is especially important when we deal with people, agencies, and companies with little or no background in science or engineering. We also will try to show how ecological restoration and related disciplines can join forces to become effective agents for environmental change under the broad concept of *restoring natural capital* (RNC). In this context, we will introduce *sustainability science*, which serves as a philosophical, scientific, transdisciplinary underpinning for ecological restoration and RNC.

## Restoration Ecology and Ecological Restoration

The *SER Primer on Ecological Restoration* (SER 2004) states that "ecological restoration is the practice of restoring ecosystems as performed by practitioners at specific project sites, whereas restoration ecology is the science upon which the practice is based. Restoration ecology ideally provides clear concepts, models, methodologies and tools for practitioners in support of their practice" (SER 2004:11). This distinction between ecological restoration and restoration ecology is clear and straightforward. Ecology is the study of the relationships of organisms to each other and their environment; whereas restoration ecology is the branch of ecology that has reference to, and tries to advance the practice of, restoration. It would make no sense to use the phrase restoration ecology to signify how a restoration project is conceived, designed, and implemented; that would not be the study of anything per se but rather the application of knowledge.

Knowledge derived from studies in restoration ecology is broadly applicable to a range of applied disciplines that are described in this chapter and not just to ecological restoration. Paradoxically, restoration ecology serves a wider array of environmental fields than the discipline of ecological restoration with which it was initially intended to complement

As odd as it seems, professional researchers and teachers of restoration ecology, tend to work independently of practitioners in ecological restoration (Cabin et al. 2010). Part of the problem is that the new field of restoration ecology did not attract interest from most theoretical ecologists until the turn of the twenty-first century. Restoration ecology is simply too new a discipline for much collaboration to have developed. Prior to the advent of restoration ecology, restoration practitioners developed their own restoration knowledge (chap. 8). Researchers carefully control their experimental conditions to generate results that are acceptable for publication in peer-reviewed academic journals. They are hesitant to consider collaboration with practitioners whose restoration project sites are generally too heterogeneous and may have too many activities happening simultaneously for controlled research protocols to be observed. However, techniques and protocols that were developed and applied at project sites by restoration practitioners have sometimes been validated by restoration ecologists who tested them during the course of experimental investigations to address other, more theoretical, questions.

Project work by practitioners remains largely invisible to the research community, because practitioners rarely publish, or if they do, they don't publish in the peer-reviewed journals that researchers tend to read and cite. Consequently, many restoration ecologists remain unaware of the accomplishments of practitioners and of restoration projects that they might access for collaborative efforts. Practitioners can't be faulted, because, they are not paid or promoted on account of their publications record. Much research generated by restoration ecologists is too theoretical or specific for resolving the pragmatic problems that practitioners face. The gist of these differences between ecological restoration practitioners and restoration ecologists has been deftly captured by Robert Cabin (2011), in his book *Intelligent Tinkering: Bridging the Gap between Science and Practice*, which contrasts the approaches, constraints, and advances of institutional research and community-based restoration practice in Hawaii.

As stated in chapter 8, we propose that this gap only represents the temporary growing pains of a new discipline. It is already beginning to close as ecological restoration becomes more accepted as a mainstream discipline. One mechanism that will help precipitate closure will be the pending launch of a professional *certification* program in ecological restoration by the Society for Ecological Restoration (SER) described in chapter 11. This program will attract both practitioners and researchers and will require those who are certified to be acquainted with both fields.

Restoration ecologists have worked and published extensively in recent years on ecological theories such as succession theory, community assembly rules, state-transition models, and considerations of novel ecosystems. This output provides perspectives and good background reading for practitioners, but it has little direct application at project sites. Instead, practitioners are in need of studies in *ecophysiology*, hydrology, geomorphology, and autecology, and they need assistance with regional inventories of intact ecosystems from which reference models can be prepared. These are the sorts of studies that practitioners can apply directly into holistic ecological restoration projects with the intent of restarting impaired ecological processes and reestablishing historic continuity.

## Ecosystem Management

Another area of recurrent confusion is the distinction between restoration and ecosystem *management*. According to Edward Grumbine (1994), ecosystem management attempts to maintain ecological integrity at various levels of organization (e.g., species, populations, communities, and ecosystems) and to maintain evolutionary and ecological processes of ecosystems. Ecosystem managers manipulate ecosystems to compensate for modern human impacts on the environment, which, if not addressed, would degrade ecosystems to the point that they will need restoration.

Some ecosystem managers conflate routine ecosystem management with ecological restoration. Such management consists of prescribed burning or the application of other stressors that prevent intact ecosystems from degrading. Our position is that the roles and objectives of management and restoration differ, and the distinction between them should be respected. In chapter 1, we proposed that ecological restoration recovers an impaired ecosystem until it has displayed the capacity for self-organization and therefore the potential for sustainability. At that point, the ecosystem has been restored, and thereafter, ecological management compensates for subsequent anthropogenic influences that may interfere with the capacity of the restored ecosystem to continue its self-organization.

Exactly the same kinds of manipulations that were needed to restore an ecosystem may have to be continued as ecosystem management after all restoration project tasks have been completed. Prescribed fire is a prime example in instances where burning was a principal technique to accomplish restoration. After restoration, periodic prescribed fires will be needed into the indefinite future—as part of ecosystem management—to remove plants of competitive, fire-sensitive species that had colonized since the preceding fire. Removal of ruminants or control of their grazing is another possible restoration task that may have to be continued later as ecosystem management. Let us be clear, though, that ecosystem management is not the purview of ecological restoration. Once ecological restoration tasks are fully discharged, the management of the restored ecosystem become the responsibility of other parties, or, alternatively, the restoration practitioner changes roles, puts on a different cap, and becomes an ecosystem manager.

Parties that administer a project site following the completion of restoration tasks can do with it what they please, just as patients, once healed by a physician, may resume their lives as they please after they are discharged from the clinic. The ideal option for administrators of a formerly restored ecosystem would be to provide protection and any needed ecosystem management to allow its open-ended development as a natural area, particularly since long-term self-sustainability was the ultimate rationale for conducting ecological restoration in the first place. However, other options are available. One option would be to manage the ecosystem in a way that ensured that it eventually attained its target or reference state, which would mean that intentional design was imposed and the naturalness of that ecosystem was sacrificed accordingly. Another option could be to dedicate the site for the extraction of timber or other resources. Neither of these options pertain directly to ecological restoration, and electing them would seem to represent overexpenditure in terms of restoration effort for the narrowly defined benefits that would accrue.

Those who prefer not to distinguish these two fields frequently claim that you can't tell when an impaired ecosystem has recovered to the point of self-organization and self-sustainability. This concern is equivalent to stating that you can't be sure if your patient is well enough to be discharged from the clinic, or if your daughter is mature enough to start dating. These are all risks that cannot be avoided and require judgment based on experience. A major reason for including time for aftercare in project plans (chap. 9) is to give practitioners, or anyone else with authority for project oversight, ample opportunity to decide when a restoration project can be closed. If the ecosystem later proves not to be self-organizing, then more ecological restoration can be applied, just as the patient can return to the clinic for additional treatment.

## Rehabilitation, Reclamation, Revegetation, and Remediation

Those who are engaged in ecological restoration need to be able to distinguish it from related disciplines and to know how those other disciplines can be brought into play effectively to solve environmental issues. In particular, we refer to rehabilitation, reclamation, revegetation, and remediation.

*Rehabilitation*, according to the *SER Primer*, is a term that describes the reparation of ecosystem processes, productivity, and services rendered without regard to achieving the fullest possible reestablishment of preexisting biota in terms of its species composition and community structure (SER 2004, 12). The emphasis is usually on productivity. The underlying assumption is that former functionality and ecosystem services can be recovered by substituting—or accepting prior substitutions—of other species for those that occurred in the past. This approach is often applied in landscapes or wetlands where people are living and working the land, notably in Europe and Australia. However, in principle, ecological rehabilitation shares with restoration the use of *reference sites* for the orientation of interventions to halt degradation and recover ecosystem services (Aronson et al. 1993a). It should not be confused with reallocation of a site, which means simply assigning a new use that has nothing to do with historical continuity. The European approach relies less on a historical and ecologically mature reference system than in North America (Moreira et al. 2006), probably because Europe lacks what is considered to be "original" predisturbance ecosystems. All ecosystems and landscapes have been altered substantially and repeatedly by human activities, and reference sites are necessarily semicultural ecosystems.

Rehabilitation has been the term of preference to designate ecological improvements of grasslands, arid rangelands, and grassy woodlands that are prevalent in Australia (Noble et al. 1997), particularly by those who apply a regeneration model to land management, as opposed to a historic reconstruction model (McDonald 2005; Tongway and Ludwig 2011). The regeneration model uses different grazing regimes and fire to encourage native range grasses and discourage encroachment by undesirable shrubs and exotic plant species, thereby improving function (Prober and Thiele 2005). Historic reconstruction is hampered by a lack of reference sites that have escaped two centuries of livestock grazing by ranchers whose ancestors emigrated from Europe. The original grassland environment was quite heterogeneous (Noble et al. 1997). The intentional return of habitat heterogeneity, coupled with control of invasive, nonnative species, judicious use of fire, and modest grazing pressure, is a reasonable strategy for accomplishing ecological restoration. In other words, much rehabilitation in Australia aims to reestablish historic continuity and thus readily qualifies as ecological restoration. The Australian example illustrates that ecological restoration and rehabilitation are closely allied and partially overlapping disciplines that both recognize the functional role of biodiversity that has evolved at a given site in determining how an ecosystem works and evolves over time.

The definition of ecological restoration adopted by SER (2004) and restated at the beginning of chapter 1 encompasses much project work that has been called rehabilitation by respected investigators such as Lamb and Gilmour (2003). This book adopts and further clarifies that position; however, some projects remain that fail to qualify as restoration because of their real or assumed disconnection from a historic trajectory and from a reference model that reestablishes historic continuity. Such work recovers ecological processes for the express purpose of increasing the flow and reliability of multiple ecosystem services to people. Therefore it should be considered as rehabilitation. In Spanish-speaking and Portuguese-speaking countries, especially in Latin America, a prevalent term for the kinds of projects discussed here is *recuperación*. The meaning in Spanish and Portuguese is very similar to rehabilitation in the sense we employ that term. Rehabilitation has been a favored term by those who work on aquatic ecosystems, probably because the composition and structure of submerged ecosystems are commonly obscured from view and consist of nonsedentary organisms.

*Reclamation* is an older term to designate conversion of land, wetland or shallow seas, which are perceived as being economically worthless, to a productive condition, commonly for agriculture, aquaculture, or *silviculture*. Recovery of productivity is generally the main goal. However, the original sense of the word, re-*claim*-a-tion, implies taking or retaking something from nature. In the Netherlands, for example, seawalls were built along the entire coastline in the thirteenth century to convert shallow estuaries and tidelands for pastures and other agricultural production—a practice that actually started in the seventh century BC (Bakker and Piersma 2006). The practice was exported elsewhere to reclaim natural tidelands and freshwater wetlands, which were drained, filled, or protected with dikes from tidal impacts. Currently, this process is being reversed in the Netherlands, for example, in the Dutch portion of the Wadden Sea (Aronson and Van Andel 2012), England, and elsewhere in order to restore wetlands for "nature development" and to reduce pollution from agrichemicals.

To take another example, the pits created by opencast surface mining are typically reclaimed by backfilling with overburden spoils or with tailings consisting of uneconomic materials after ores were separated from them. The reclaimed land is then available to be reallocated for a new use. Reclamation laws in the United States and some other countries require that backfilled and physically reclaimed land be stabilized and vegetated with grasses and legumes or another cover, which may consist of only a single species of plant that may or may not be native. Regard for public safety and health are also factors in reclamation designs, but without regard to ecological conditions informed by a reference site. Therefore it is erroneous to consider reclamation of this sort to be a synonym for rehabilitation, as indeed is often done. A physically reclaimed mine can be rehabilitated or even restored (e.g., Clewell 1999; Gardner and Bell 2007), but mining companies typically do only what is required, which seldom exceeds simple revegetation.

To add further to the confusion, the term reclamation is sometimes applied to the creation of wetlands and ponds in an urban or industrial context for the storage of water or its reconditioning for reuse. Imprecise use of language of this sort is rampant in the emerging markets of carbon credits and carbon trading. Readers should be aware of the ambiguities of such technical shorthand and technocratic jargon and eschew their use in preference for accurate terminology.

To go one step further, much intentional greening of degraded or physically modified reclaimed land is often called restoration, even though the terms *revegetation* or—when trees are used—*reforestation* would be more precise. Narrow road embankments are sometimes suitable environments for the establishment of native plant communities, which provide welcome aesthetic relief for motorists and some other ecosystem services such as shade. Installations of such communities have been equated with ecological restoration; however, spatial constraints largely limit their status to that of ecotones. Their trophic structure is necessarily limited on account of highway safety issues. Nonetheless, activities on such sites that approach restoration are welcome alternatives to traditional roadside maintenance.

Both reforestation and revegetation are gaining more precise definitions in the international treaties forming part of the UN Framework of the Convention on Climate Change; to wit, reforestation is defined as "the direct human-induced conversion of non-forested land to forested land through planting, seedling and/or the human-induced promotion of natural seed sources, on land that was forested but that has been converted to non-forested land" (UNFCCC 2001). Similarly, revegetation is defined by the UNFCCC, in an unusually narrow fashion, as "a direct human-induced activity to increase carbon stocks on sites through the establishment of vegetation that covers a minimum area of 0.05 hectares and does not meet the definitions of afforestation and reforestation contained here" (UNFCCC 2001). These definitions allow the planting of nonnative species.

*Remediation* is a term that refers to the reduction or elimination of contaminants from a place where they are not wanted. This is a task that could be included during site preparation in those ecological restoration projects that occur on contaminated sites, such as oil spill areas. There are two kinds of remediation. Phytoremediation is the process of removing toxic metals or other substances from the soil or substrate, using plant species that are known to accumulate the substances in question in their tissues. These plants are harvested and removed to prevent metals or other unwanted substances from returning to the soil as part of the humus. Bioremediation is the term describing a large suite of techniques that help decontaminate a site. It refers to the process of introducing bacteria that metabolize petroleum, of which a number of natural species have been identified, isolated, and grown in quantity for application on soils or in waters containing oil spills. Other techniques employing microorganisms can remove excess nitrates, foul odors, and other contaminants from soils and bodies of water.

## Compensatory Mitigation

*Mitigation* is the term used by public agencies in the United States and a growing number of other countries to describe an act of compensation undertaken as an offset for expected or incurred losses of biodiversity or other environmental values resulting from activities in support of economic development and improvements in infrastructure. The full term is compensatory mitigation, and in the United States, at present, it may be legally satisfied by revegetation, remediation, reclamation, rehabilitation, restoration, or other kinds of project work, as prescribed in permits that authorize private development or public works activities. It is common usage to say that a site has been mitigated. However, the term mitigation should not be used to refer to work that is conducted at a project site because mitigation is an administrative or legal option and not an onsite intervention. The emphasis is on the legal process rather than on the endpoint reached (Bradshaw 1996). In the manner that mitigation is currently used, compensation does not mitigate or lessen the impact. Instead, it lessens the effects of environmental damage, commonly after the fact. Compensatory mitigation is sometimes used to refer to the long-term preservation of ecologically valuable lands that would otherwise be subject to development, which does not lessen the severity of the impacts that were allowed to happen. Again, this is a misleading use of technical jargon.

It stands to reason that ecological restoration, as we have described it, should be performed as mitigation in many circumstances, but that unfortunately is rarely what happens. When a mitigation activity is called restoration, it is rarely conducted in a holistic sense whereby practitioners make use of a reference model and seek to reestablish historic continuity. Economic interests assert sufficient influence over political processes and policy development to ensure that less expensive and less time-consuming activities are mandated as mitigation requirements by public agencies.

In negotiations of offsets, very often there is a concern for how much is enough, and what kinds and degrees of environmental improvement would be equivalent to the environmental impact that is permitted, or that has occurred accidentally. Habitat Equivalency Analysis (HEA) is one of various assessment methods that have been proposed (Quétier and Lavorel 2011), and are being tested, to resolve habitat equivalency issues. The idea is simple: if you destroy a certain area of a healthy, functional wetland or other ecologically sensitive system, a functionally equivalent area is replaced, be it on land or at sea (Price et al. 2012). To further explore these questions, a clear distinction must be drawn between ecological restoration on the one hand, and fabrication or creation, on the other.

## Fabrication and Creation

*Fabrication* is the replacement of a previously occurring ecosystem with an entirely different kind of local ecosystem following a radical change in physical site conditions. *Creation* is an alternative term for fabrication, frequently used to describe the installation of a wetland ecosystem on a scraped-down upland to satisfy requirements for compensatory mitigation. Fabrication is exemplified by the intentional establishment of tidal marshes and dune ecosystems on islands created by deposition of dredge spoils along riverbanks and as new islands in estuaries. Deposition of dredge spoils creates a site with new and radically different physical site conditions that can only support a contrasting kind of ecosystem relative to aquatic systems that occurred previously. Its installation initiates a new ecological trajectory. No historical continuity was reestablished, and ecological restoration was not accomplished, because there was nothing recovered and thus nothing restored.

Drastically modified environments may create physical conditions that are not represented locally, and the ecosystem that is installed may deviate in terms of species composition, structure, and even ecological processes from those in the region. For example, dredge spoil islands may consist largely of silts and clays that were removed from bottom sediments, rather than sands that characterize naturally occurring islands that serve as reference sites. Restoration of mine land where a contrasting kind of ecosystem is substituted for the one that occurred prior to mining would also qualify as fabrication. Although fabrication and creation are not ecological restoration, these activities are certainly welcome in some situations as environmental improvements, so long as they are not used as an excuse for wanton destruction of remaining natural ecosystems. Furthermore, in practice, the fabrication or creation of habitat for a single species of favored plant or animal is all that is sought, and all that is achieved. Still this may be much better than nothing, in the way of compensation for authorized environmental damage or fragmentation.

We cannot leave the topic of fabrication without mentioning a truly fabulous project from Salo, Finland. Decommissioned sewage treatment lagoons, known as Halikonlahti Bird Pools, support 110 species of breeding birds whose eggs and chicks are prey to small mammals (fig. 10.1). To protect nests and treat chronic water quality issues, ecological artist Jackie Brookner built floating islands from plastic tubing and mesh. She planted native wetland species on them, and their roots extended through the mesh and into the water column. Microbial biofilms coated the roots and served as the main agents of filtration for purposes of phytoremediation. Birds nested unmolested among these plants and were shaded by sculpted lightweight artificial rocks. The project is a floating work of art called Veden Taika or the Magic of Water and has generated much civic interest and pride. The project qualifies as fabrication, because it does not reestablish historic continuity and can hardly be called sustainable, but it is informed by ecological restoration and demonstrates the expanding influence of ecological restoration into related fields, including ecological art for which Veden Taika serves as an outstanding example.

## Landscape Architecture and Design

Professional training and experience in landscape architecture and design is a preferred path for many professionals to move into ecological restoration. The two disciplines share many similarities. When landscape or ecosystem design is woven

**FIGURE 10.1.** An unusually inventive example of ecosystem fabrication by Jackie Brookner, consisting of a floating island constructed of plastic and artificial rocks to support real plants and real birds in Salo, Finland.

with nature, there is much room for synergy and collaboration with ecological restoration, with one caveat pertaining to the meaning of "design." In landscape architecture, design commonly implies that the system will be manipulated until the original target or desired endpoint is reached. Ecological restoration is open ended, as we noted in chapter 9. Design has no connection to the endpoint in ecological restoration, and we avoid using the design to denote specific methods and details in project plans. Another distinction with landscape architecture concerns aesthetics as an intentional product of artifice. In ecological restoration, aesthetics is an emerging property arising from natural processes during ecosystem recovery rather than a specifically designed outcome.

## Ecological Engineering

The *SER Primer* (2004, 12), states that "ecological engineering involves manipulation of natural materials, living organisms and the physical-chemical environment to achieve specific human goals and solve technical problems. It thus differs from civil engineering, which relies on human-made materials such as steel and concrete." The evolving relationship between ecological restoration and ecological engineering remains somewhat murky in academia, where these sister disciplines sometimes appear as quarreling siblings. However, in real-world situations, practitioners are working in both fields simultaneously and making substantial contributions toward ecological improvement and *human well-being*.

Ecological engineering was first proposed a generation ago by the innovative and highly influential ecologist Howard T. Odum (1924–2002). He contended that many problems that are ordinarily resolved by civil engineering with the use of inert materials could be solved effectively and much less expensively by use of biological materials—organisms and their detritus—in a manner that relied on ecological principles and processes. Odum was personally engaged in developing ways to treat and recycle wastewater—mainly sewage effluent—in wetlands to remove suspended solids, excess nutrients, infectious organisms, and contaminants. This application of ecological engineering remains its crowning achievement and will only grow in importance as potable water supplies continue to decline globally.

The field of ecological engineering developed concurrently with the field of ecological restoration. Its principal membership association is the International Society of Ecological Engineering, which publishes the journal *Ecological Engineering*. Former students of H. T. Odum have been in the forefront of its development, notably wetland ecologist William Mitsch of Ohio State University. The field has not been embraced unconditionally by traditional engineering fields and by civil engineers, whose quest for mathematical precision and long-term predictability of their output cannot easily accept the "messy" introduction of living organisms by ecologists.

Ecological engineers have responded by going to great lengths to demonstrate that ecological science is as well founded and rigorous as that of traditional engineering. For example, Mitsch and Sven Jørgensen (2004), in their important book *Ecological Engineering and Ecosystem Restoration*, made the point repeatedly that ecological engineering is predicated on principles of systems ecology. This proposition may be true for wastewater treatment but not for most applications of ecological engineering that were enumerated by Patrick Kangas (2004) in his equally valuable book *Ecological Engineering: Principles and Practice*. Most of these were agronomic or silvicultural applications (soil bioengineering, bioremediation, phytoremediation, compost engineering), bioassay techniques (ecotoxicology), and advanced forms of food production (aquaculture, hydroponics). Kangas (2004) also listed wastewater treatment, wetland mitigation, and the reclamation of disturbed lands as applications of ecological engineering. These are more readily identifiable with systems ecology, but they are not necessarily informed by local reference sites or designed to become self-sustaining ecosystems.

Some and perhaps many ecosystems designed or constructed under the aegis of ecological engineering are built to specifications that will facilitate or maximize ecological processes as the solution to a particular problem. Such projects may include constructed civil engineering features (e.g., dikes, weirs, drainage tiles, culverts, pumps). They may also need external subsidies of energy and materials, as would be typical of production systems. The installation of such ecosystems ordinarily would not qualify as ecological restoration in terms of reestablishing historic continuity and reinitiating ecological processes as informed by a reference model. This in no way detracts from their usefulness and desirability, and we see many opportunities where engineered ecosystems can exist side by side with restored ecosystems in well-functioning semicultural landscapes. For example, water discharged from land dedicated to agriculture could pass through a constructed wastewater treatment wetland for removal of contaminants and excess nutrients before entering a restored ecosystem that conserves biodiversity and provides ecosystem services. By contrast, constructing a "green" roof for a building in a city falls within the domain of ecological engineers and gardeners. We note that in many countries, the terms environmental engineering or landscape engineering are used to mean the same thing as ecological engineering. Another relevant term is industrial ecology, which in general refers to systems created to manage industrial or urban wastes.

When predictability of the endpoint is not at issue, the scope of many ecological engineering projects could be expanded until they qualify as restoration. Conversely, some authors argue that ecological restoration is an element of ecological engineering or that the two terms are synonymous or nearly so. Both Kangas (2004) and Mitsch and Jørgensen (2004) considered ecological restoration to be a subset of ecological engineering and not an independent discipline. We argue just the opposite, as noted at the beginning of this chapter. In ecological restoration projects or programs, engineering activities of many kinds can serve as tools or components to advance the project and achieve the overall goals. We therefore strongly advocate that ecological restoration be recognized by and evaluated according to the criteria stated in this book, including the use of a reference model and the reestablishment of historic continuity. In particular, we would not want ecological restoration to be identified or confused with the production of *designer ecosystems* or custom-built installations that are constructed to fulfill narrowly conceived or short-term societal needs, such as green roofs, roadside revegetation, or wastewater treatment. Happily, the field of ecological engineering—and landscape architecture and design—are evolving just as rapidly as ecological restoration. Proponents of all three fields are increasingly recognizing the realities of nonequilibrium and nonlinear dynamics in the ecosystems with which they are concerned. The notion, first proposed by H. T. Odum, of using "ecological engineers" consisting of nonhuman species (Jones et al. 1994) as helpers and role models (Rosemund and Anderson 2003) in ecological design and restoration is also encouraging and promising, even though confusion persists in the use of this term *and* ecological engineering as defined herein.

In closing this section, we note that the term ecological engineering is actually a metaphor. It is not just engineering; it is also applied ecology used to perform practical services for people in a manner that in many ways is superior to, and less expensive than, traditional engineering. Ecological engineers are simply offering a better service to the public than traditional engineers, and the latter should seek training in ecology if they are to remain competitive, just as landscape architects should study botany and ecology, as well as landscape history and design. Similarly, ecological restoration is a metaphor that seems to capture the imagination of people of all cultures (Munro 2006). Ecological restorationists do not *actually* restore ecosystems; instead they restart, revitalize, reorient, or accelerate inherent ecological processes. They recover lost services, and recuperate or even augment natural capital, a concept developed in the next section.

## Restoration of Natural Capital

*Natural capital* is a term from economics that refers to natural resources on which people's well-being depends. In the context of ecological restoration, these resources consist primarily of natural and semicultural ecosystems, which, under normal conditions, are not depleted as they provide continuing flows of natural goods and services. Natural capital also extends to production systems, lands dedicated to economic infrastructure (e.g., powerline rights-of-way), and abandoned lands that are vegetated or retain enough biotic content to provide at least some flows of natural goods and services of benefit to people. The restoration of natural capital (RNC) refers to ecological restoration or rehabilitation, and to any activities within the broad purview of ecological engineering, which increase flows of natural goods and services from ecosystems, production systems, and other forms of natural capital that suffered degradation, if not outright impairment. In addition, the RNC concept embraces social capacitation, so that stakeholders, the community, and local institutions understand and appreciate natural capital and its benefits, and how their own natural capital can be restored and managed in a sustainable manner.

Following Aronson et al. (2007a), we define the restoration of natural capital succinctly as the replenishment of natural capital stocks in the interests of long-term human well-being and ecosystem health. *Stock* refers to a specific unit or quantity of economic capital, as, for example, a particular ecosystem. The use of stock in an RNC context is economic recognition of the maxim that Nature sustains us.

Ecological economists, some restorationists, and, increasingly, some journalists and the media, combine biodiversity, ecosystems, and the renewable and nonrenewable natural resources they contain under the stimulating term natural capital, often without seeking a full understanding of the term. The first step is to distinguish between the stocks of natural capital and the *flows* of ecosystem goods and services. Jurdant et al. (1977) initially used the term natural capital in a report to the Quebec regional government, and Costanza and Daly (1992) introduced the concept to a large academic readership that gradually embraced it. The term RNC was suggested by Cairns (1993) and introduced to ecological restorationists by Clewell (2000b). The concept was developed by Milton et al. (2003), Aronson, Clewell, et al. (2006), Aronson, Milton, et al. (2006), and Aronson et al. (2007a). In this approach, ecological restoration becomes a strategy for sustainable economic development, as well as a nature conservation strategy (Blignaut et al. 2008, 2011). Ecological restoration and RNC are bridges that reconcile legitimate long-term economic development goals with those related to nature conservation. Adding the concept of RNC to our vocabulary along with ecological restoration sensu SER 2004, can help find common ground and encourage cooperation between ecologists, environmental lawyers, politicians, and economists as they address the interrelated problems we face today. Needless to say, causes of degradation must be addressed, as well as the symptoms to achieve long-lasting restoration (Blignaut 2008). Figure 10.2 shows that process in action, as experts and officials meet to identify causes of degradation and prepare RNC plans to eliminate them.

Ecological economists and proponents of RNC generally call on society to invest not only in conservation but also in the restoration or *augmentation* of the stocks and reserves of fundamental *assets*—natural capital—on which all human societies and economies depend (TEEB 2010). The flows of natural goods and services that accrue from these stocks of natural capital are equivalent to the dividends that accumulate from prudently managed financial capital. As shorthand, economists lump natural goods as one form of natural services—provisioning— and designate them all under the term ecosystem services.

Production systems can be distinguished as cultivated capital, whereas natural and semicultural ecosystems, along with their native biodiversity, are termed renewable natural capital. When speaking of restoring natural capital, we refer to both renewable and cultivated natural capital, but not to nonrenewable natural capital such as diamonds, gold, and petroleum. We also emphasize that biodiversity is not an ecosystem service but rather part and parcel of natural capital. This point is very often misunderstood, with the result of muddy thinking on this broad new area of cross-disciplinary thought and action.

Programs to restore natural capital (Aronson et al. 2007a, 2007b) at the land-

**FIGURE 10.2.** Field trip to plan an RNC program to improve potable water quality and biodiversity in Ecuador. A. Clewell (3rd from left), J. Aronson (3rd from right), and Australian Bev Debrincat (center) meet with local technical personnel to develop strategies and tactics.

scape level include the ecological restoration and/or rehabilitation of ecosystems, ecologically sound improvements to production systems, ecologically sound improvements in the utilization of biological resources, and efforts to increase public awareness and appreciation for the importance of natural capital.

RNC includes a range of core concepts and activities, as follows:

- Recognizing natural and semicultural systems as stocks of natural capital, including natural and semicultural ecosystems and production systems devoted to agriculture, aquaculture, silviculture, and the like.
- Augmenting stocks of natural capital through ecological restoration in a holistic sense and rehabilitation that includes rehabilitating degraded production lands and waters, so as to improve their utility and reduce their negative impacts. Goals include preventing soil erosion and pollution of bodies of water; eliminating soil compaction and increasing soil organic matter; establishing cover crops and nitrogen-fixing vegetation; eliminating the causes of eutrophication, desertification, and salinization; and implementing other appropriate technologies offered by environmental engineering such as phytoremediation, and bioremediation.
- Reintegrating fragmented landscapes in order to conserve biodiversity (e.g.,

corridors to connect existing conservation areas, and set-aside nature reserves) and to improve landscape resilience and sustainability.

- Creating ecologically sound designer ecosystems, for example, "living" roofs and city parks, and living systems for roadside repair and wastewater treatment, or fostering the development of rehabilitated ecosystems that serve human needs in locations where historic continuity cannot be reestablished.
- Planning for and encouraging the implementation of best management practices to protect and maintain natural capital stocks and to augment flows of natural goods and services. This applies, for example, to fishing, mining, and other systems of exploitation, transport, and distribution of goods and services. It can embrace organic and biodynamic farming and the markets its produce attracts. Often it will require technologies offered by environmental engineering, phytoremediation, and bioremediation.
- Restoring relevant social capital in order to increase public awareness of flows of natural goods and services. Armed with such knowledge, individuals and their local associations and institutions are motivated to engage in RNC and in the protection and management of natural capital. Such engagement, in turn, promotes the equitable and sustainable distribution of natural goods and services.

RNC is thus a much broader concept than ecological restoration; it incorporates all investments in natural capital stocks—especially renewable and cultivated—in ways that will improve the services of both natural and human-managed ecosystems within landscapes while contributing to the socioeconomic well-being of people. It entails reducing avoidable losses of natural capital to pollution, short-term exploitation of resources, and unregulated development. It includes publicity and educational programs to raise public awareness on the benefits and importance of natural capital in regard to everyone's well-being.

To date, ecological restoration is not appreciated or financed in less affluent parts of the world unless it benefits people directly. For that reason, restoration practitioners in those regions will have no choice but to approach their craft from the perspective of RNC. Academic programs that train practitioners should consider technical and conceptual instruction and hands-on training in RNC.

Some advocates of ecological restoration who are motivated by a biotic rationale, as explained by Clewell and Aronson (2006), and whose attention lies with the perpetuation of biodiversity may raise a concern here. They may argue that RNC's human-centered focus obscures the ethical proposition that ecosystems and all the processes and species they contain are worth restoring and preserving for their own sake, regardless of their economic or cultural value to people. We acknowledge this difference in motivation, but we note that RNC reaches the same conclusion: that all of the processes and species of ecosystems are worth preserving.

## Sustainability Science

*Sustainability science* is the study of the dynamic interactions between nature and society. The ultimate goal of sustainability science is to create and apply knowledge in support of decision making for sustainable development that is socially just for present and future generations. The application of research in sustainability science is to resolve specific problems related to natural resource utilization in an appropriate and effective manner (Kates et al. 2001; Clark and Dickson 2003; Weinstein and Turner 2012).

Like the terms mitigation, reclamation, ecosystem services, and carbon trading, sustainability can be interpreted in many ways, and a great deal has been written on the subject of its use and misuse. Yet it is a powerful word that is worth maintaining. An enormous step forward toward human well-being and world peace would be for local and national economies everywhere to cease pursuing perpetual economic growth that merely consists of converting renewable and nonrenewable natural capital into manufactured capital and financial capital, with no consideration for offsets or durability, and insufficient effort to avoid damage, or to repair damage to our natural capital, which provides the underpinnings of society and serves as our "ecological infrastructure." Instead of promoting unsustainable growth, which Daly and Cobb (1994) famously called "uneconomic growth," we should instead manage economies to become stabilized and ecologically sustainable for subsequent generations (Costanza and Daly 1992; Daly and Farley 2004). This vision requires the conservation and wise use of natural resources and a substantial admixture of active investment in RNC. As suggested in the previous section, the restoration of impaired ecosystems and production systems will increase the flows of goods and services for people and economies while creating jobs, livelihoods, and increased social capital. The benefits far outweigh the costs when viewed from an intergenerational perspective. Obstacles to this paradigm and policy change are gigantic and largely irrational. Ecological restoration and, more broadly, RNC, deserve recognition as central strategies for human well-being and the search for a sustainable *and* a desirable future.

# Chapter 11

# Projects and the Professional

In this chapter we identify stakeholders and project sponsors as principals in restoration projects. Then we identify the roles and describe the responsibilities of personnel who participate in ecological restoration projects on behalf of the sponsor. We continue with descriptions on how restoration projects are organized and administered. Finally, we describe the knowledge base and breadth of experience that competent professionals in ecological restoration share, a body of expertise that promises to become the basis for professional certification.

## Stakeholders

We begin with *stakeholders*, because they are those persons and organizational entities who are most likely to be affected by a restoration project and who would be its principal beneficiaries. They are called stakeholders because they have a personal, cultural, or economic stake in the project. Stakeholders ask if the project will fulfill their values, as described in chapter 2, or if it will produce negative consequences. Stakeholders may be local or absentee, depending upon their proximity to the restoration project site and how directly they are affected by it. Proximity is defined on a case-by-case basis. Absentee stakeholders may own property in the proximity of the project but reside elsewhere. Other absentee stakeholders may contribute to a philanthropic organization that finances restoration projects. For example, donors from an affluent nation may specify that the philanthropic foundation to which they donate shall underwrite the restoration of tropical rainforest in a nation that they may never visit. In return, they will have the satisfaction of doing their part to support the biosphere and its biodiversity.

Stakeholder organizations can be public or private, for-profit or not-for-profit.

Commonly, stakeholders can be divided along three lines. One consists of local individuals, along with community-based organizations (CBOs) and local institutions, both private and public. Those persons who are associated with these local interests are usually united by shared cultural values. The second consists of economic entities such as corporations and agricultural interests whose resource base could be affected by a proposed ecological restoration project. Increasingly, corporations in extractive, energy, transport, or other industries, are obliged to undertake restoration as an offset for unintended or unavoidable environmental damage. Some industries are beginning to be self-governing in this respect, even in the absence of legal constraints or incentives. An example is the forest products firm that is participating in the Chilean restoration described in VFT 7. In all cases, the corporations at issue have a clear stake in the successful outcome of restoration projects.

The third group of stakeholders consists of governmental agencies with responsibilities for protecting and allocating natural resources. Such agencies are generally obligated to protect the welfare of people and determine the efficacy of a proposal on the basis of the greatest and highest good. A growing number of international treaties and conventions are reinforcing the need for all nations to invest far more in the conservation and restoration of our limited natural capital. Some stakeholders, however, are all too commonly ignored in the development of larger ecological restoration programs that are administered in a top-down fashion by an external authority or large corporation. We argue that local stakeholders, particularly, should have a strong voice in the decision of whether or not to initiate an ecological restoration project and how it should be conceived.

Stakeholders may of course disagree as to the prudence of an ecological restoration project. Such disagreement led to angry protests in suburban Chicago, when many residents served as volunteers to restore tallgrass prairie and their neighbors, especially animal rights activists, invoked strong political pressure to prevent restoration from continuing (Shore 1997). We advise sponsors of potentially controversial restoration projects to engage professionally trained social scientists as liaison officers to identify stakeholders, listen to their views, and to try to negotiate consensus among them. Such negotiations raise ethical issues whereby democratic ideals are balanced with socioeconomic expediency or, as in Chicago, opposing cultural values must be reconciled. Transparency and disclosure, coupled with effective publicity and serious journalism, are essential to defuse such problems before hardened positions develop.

## Project Sponsors

The project sponsor can be an institution, organization, or any other entity that assumes overall responsibility for a restoration project; secures funding for it; and assembles professionals who plan and implement it. A restoration project is identified, at least in part, by the organization that sponsors it. The sponsor may be an agency of government at any level—local, state or provincial, regional, or national. Sometimes public agencies are required by law to sponsor ecological restoration as compensatory mitigation, particular highway departments and other agencies responsible for public works projects that impinge on wetlands and other ecologically sensitive areas. Sponsors may be transnational organizations such as the UN Environmental Programme, the European Union, the World Bank, regional development banks, or nongovernmental organizations (NGOs) that operate internationally, such as the World Wildlife Fund, The Nature Conservancy, and the Wildlife Conservation Society. Sponsors may also be a for-profit corporation, particularly those that are legally obligated to perform restoration as mitigation or to earn environmental certification for extractive products such as wood (VFT 7). Other sponsors may be NGOs that operate at a local or national level. A sponsor could be a philanthropic foundation; school, university, or research institute; a public museum, arboretum, botanical garden, or zoological park; a professional association; a branch of the military; a monastery or other religious order; a tribal council of elders; a women's self-help group—which are becoming increasingly common in India and Latin America; a community-based organization of any sort; or an individual landowner. Commonly, the sponsor is a consortium of different entities, one of which assumes the principal administration of the project. One virtual field trip included in this book was sponsored by an NGO (VFT 1). The others were sponsored by consortia of public agencies (VFTs 2, 5) or collaborations between agencies and NGOs (VFTs 3, 7). VFTs 4 and 6 were sponsored by consortia of multiple organizations that prominently included universities.

The sponsor decides the administrative structure for a project and provides oversight to ensure its satisfactory completion. The project may be accomplished in-house using the sponsor's own employees or members. Some or all of the work can be delegated to outside individuals, consulting firms, workers' cooperatives, private nurseries, or other organizations under contract, subcontract, purchase order, or some other agreement to provide services. Labor can be provided by paid personnel or by volunteers who work without monetary compensation. To a restoration practitioner who is contracted, the sponsor is usually known simply as the client.

## Project Roles

Every ecological restoration project requires personnel to fulfill certain roles. These roles contribute to any of three major functions: administrative, technical, or supportive. Administrative personnel include the project director, project manager, safety officer, volunteer coordinator, and training officer. Technical personnel include restoration practitioners, biotic resource providers, equipment operators, natural scientists, planners, and social scientists. Support personnel include financiers, accountants, an office manager, attorneys, and publicists.

Organization charts may shift roles among the major functions, and they may assign titles other than those that we use to identify project personnel. In smaller projects, a single person may assume more than one role. Sometimes an individual landowner assumes all roles and performs the restoration by him or herself from start to finish. In larger projects, additional roles may be added, such as liaison officers who coordinate participation by multiple agencies, when the project is an element in a landscape-scale restoration program. Project organization becomes even more complex as contractors and subcontractors are included, with their own hierarchies of personnel and departments that share project responsibility.

## Technical Personnel

A *restoration practitioner* is someone who personally conducts or supervises ecological restoration in the field at project sites. Specifically, practitioners engage in project implementation and aftercare. In many projects, practitioners perform or participate in all tasks that occur at a project site as well as in the preparation of project plans. Onsite tasks include inventory of a project site prior to the initiation of restoration activities; selection and inventory of reference sites and the preparation of the reference model; preproject monitoring of the abiotic environment to establish baseline conditions; monitoring project sites after they have undergone restoration; and in addition, preparation of monitoring reports. A practitioner may double as the project manager. A practitioner may be employed by the sponsoring organization; engaged under contract as a consultant, contractor, or subcontractor; or engaged as a volunteer. A restoration project may be accomplished by a single practitioner, or two or more practitioners who work collectively on all aspects or separately on different aspects of a project. The chief practitioner, if one is appointed, supervises other practitioners and is responsible for the overall conduct of onsite restoration activities. A practitioner may assume broad responsibilities and authority for conducting restoration or may serve only as a technician who performs specific tasks assigned by the project manager.

The *restoration planner* (or a planning staff) prepares project plans, including maps, drawings, and written instructions as needed. Ideally, the practitioner contributes substantially to the planning process or even serves as the planner, as commonly happens on smaller projects that entail little in the way of government permits or outside contractors. The degree of detail in project plans may vary widely between projects, depending on project size and complexity and on the requirements of the sponsoring organization. Much detail may be required by government agencies and transnational organizations, whose approval is needed prior to project implementation. Project plans typically are appended to permits and are carried out as a permit condition. Detailed plans are also useful for preparing contract stipulations that are to be followed by the firm that provides practitioner services to the sponsoring organization. If contractors fail to comply with contract stipulations, monetary penalties are levied. For that reason, planning may include legal as well as technical input.

*Biotic resource providers* include seed collectors and horticulturalists that grow nursery stock for outplanting at restoration project sites. Nurseries may be established onsite, or stocks can be purchased from commercial nurseries. If animal introductions are specified in project plans, biotic resource personnel may include zoologists.

*Equipment operators* operate mechanical equipment of the sort that is used in farming and forestry operations and that is needed for site preparation activities, repair of the physical environment (e.g., filling a ditch), application of soil amendments and herbicides, and operating seed drills or other planting equipment. Restoration practitioners sometimes assume the role of equipment operators. Otherwise, skilled operators/owners of heavy equipment are frequently hired as subcontractors by the project manager.

*Natural scientists* are usually consultants who are trained in hydrology, water quality, soil science, geomorphology, and the identification of plants and animals. These are professionals who can conduct preproject evaluations and inventories of the project site and reference sites; conduct baseline monitoring; and later perform post-implementation monitoring. Experienced restoration practitioners are usually able to provide at least some of these services. Need for a *social scientist* is becoming increasingly apparent to identify and communicate with stakeholders in a liaison capacity, determine their needs and preferences, solicit their recommendations, and oversee the consensus-building process. Later in a project, the social scientist may serve as the volunteer coordinator, publicist, and organizer of public events and celebrations.

## Administrative Personnel

The *project director* is the sponsor's agent who is responsible for overseeing all aspects of an ecological restoration project. That person has a comprehensive vision for the project, including its technical, social, economic, strategic, political, historical, and other cultural aspects and implications. The project director is superior in rank to all other project personnel, including the project manager, and is responsible for the overall technical direction and leadership of a project. The project director is critically involved with the conception of a project and the development of project plans. The project director formulates or approves project goals and objectives; selects or approves reference sites and the reference model; and selects or approves the overall approach to restoration (chap. 8) and strategies and tactics for accomplishing restoration (chap. 9). The project director receives briefings from the project manager and evaluates project monitoring reports and other technical documents that are produced. The project director ensures that accountants, legal counsel, and other administrative officers understand the project and carry out their respective responsibilities. The project director represents the project before boards of directors, philanthropic foundations, funding agencies, public officials, stakeholders, and the general public. He or she may also delegate these tasks and duties to others.

The *project manager* is responsible for ensuring that a given restoration project is conducted satisfactorily on behalf of the project director and the sponsoring organization. The project manager administers day-to-day operations such as scheduling personnel, ordering and arranging for deliveries of planting stocks and equipment, ensuring adherence to contract stipulations, and approving expenditures in accordance with the project's budget. In most projects, restoration practitioners are supervised and report to a superior who is either the project manager or someone who reports to the project manager. Sometimes the practitioner does some of project management tasks, and the project manager ensures that they are accomplished. Another firm or organization that has been engaged to provide restoration services under contract sometimes appoints its own project manager. In such instances, both project managers collaborate, and practitioners receive directions from the project managers in their respective firms.

Satisfactory restoration projects require that practitioners and the project managers remain in close communication, more so than in construction projects where outcomes are more predictable. The success of many restoration projects depends on manipulating living organisms of different kinds, and the chances for surprise are great. The practitioner must react to unanticipated situations to ensure project success and cost effectiveness. Sometimes the project manager is obliged to adhere closely to schedules, budgets, and contract stipulations, which may not allow for contingencies. In such instances, practitioners should educate project managers and provide succinct information and persuasive logic that the managers can use effectively when interacting with people at higher administrative levels. We cannot overstate the importance of respectful and cordial relations between practitioners and project managers, particularly in ecological restoration projects of long duration.

*Safety* and *training officers* may be needed if personnel working onsite are inexperienced and especially if they are volunteers. Volunteerism is encouraged; particularly if dedicated volunteers are also stakeholders who may bond with a project site by participating in restoration tasks and who are likely to be the first to perceive problems or opportunities. Volunteering strengthens the importance of ecological restoration in the eyes of the public, it raises ecological literacy, and it supplies labor that may be essential for a project (figure 11.1). Some volunteers are superb workers. Others need coaching and supervision. Since volunteers are not paid, project personnel are limited in the ways they can control their participation. This is where a perceptive *volunteer coordinator* can be of great service to keep volunteers productive and happy. Some volunteers, like pet cats, can be exasperatingly independent. A volunteer coordinator plays an important role by relieving other project personnel from the time and potential hassles of communicating with volunteers, developing their work schedules, ensuring their access to the project site, providing them with tools and materials, and making

**FIGURE 11.1.** A volunteer outplants nursery-grown trees in a sheep paddock in New South Wales, Australia.

sure the experience is rewarding for everyone involved. Conversely, a golden rule when working with volunteers is to avoid wasting their time. Make sure they have important things to do, and know how to do them, whenever they come on the project site to work, or they may not return.

### Support Personnel

Support personnel handle tasks that are not directly involved in technical restoration tasks or their management, but nonetheless they are essential to the project. Restoration projects are usually expensive, and financing is central to any project. If internal funding is unavailable, external *financiers* must be identified who agree to provide project funding, and a *grant writer* may have to be engaged to attract financing.

An *office manager* is needed to keep records, coordinate administrative activities, relay communications, and produce reports. An *accountant* is needed to keep track of finances and prepare payrolls. An *attorney* may be needed to tend to regulatory permits, contractual matters, easements, property transfers, and negotiations with adjacent landowners to project sites. A *publicist* will be invaluable as a liaison with writers, journalists, photographers and the news media, particularly since technical personnel tend to be indifferent to such matters and hesitant or unwilling to interact with visitors and the news media. If a restoration project is worthwhile and executed competently, the public needs to know about it and particularly political leaders and others who develop public policy. Otherwise, there may be no interest or funding for the next restoration project or the next phase of the project at hand.

## Organizational Structure

Now we turn our attention to the contexts in which projects are administered. Basically, there are two extremes that differ primarily by the way decisions are made. One is bottom-up decision making whereby projects are initiated and their administration is maintained under local control by individuals, organizations, and institutions that are part of, and identified with, the local community and its values and interests. The other is top-down control from a central administration that is generally not local, not strongly identified with the local community, and not required to answer directly to that community. Instead, it is a regional or higher level authority. Both extremes have strengths and weaknesses. Actually, these two extremes—bottom-up and top-down—are nodes at either end of a continuum, and the organizational structure for many projects lies somewhere between them.

## Local Administration

Bottom-up projects are conceived, sponsored, planned, administered, and executed by people who reside in the vicinity of the project site. Such projects may receive outside assistance in terms of funding, expertise or other help that is locally unavailable, but the impetus and decision making is generated locally. Local projects can be those that are performed by the owner on one's own property or by a tribal village on communal land. They can be those sponsored by a community-based organization (CBO). A CBO could assume many forms—private, public, or institutional—as long as administrative and budgetary authority and responsibility for the project are locally vested. For example, an office of local government or university would qualify as a CBO, as long as that office was responsive to local concerns rather than external authority. A local restoration project could be sponsored by a local office of a large nongovernment organization (NGO) that operated local environmental stewardship programs (described later) with nominal administrative input from its parent organization. It could be administered by local citizens on public lands. Decision making is sometimes a matter of consensus by a group of people, or sometimes one person has final authority to make decisions. In the latter case, there is usually ample input from others whose opinions are valued by the person who makes the decisions. In other words, decision making is at least somewhat diffuse and may even be collegial.

In Australia, restoration is sometimes planned and conducted in national parks by local nonprofit organizations consisting of trained citizen-volunteers who are generously funded for that purpose by grants from state and federal governments. These local nonprofit organizations sometimes underwrite salaries of professionals hired by local public agencies who, in turn, assist with the administration of the restoration work. The nonprofit organizations hire local private restoration contractors as needed to perform work that is beyond the capacity of the numerous volunteers who participate in these programs. The Australian arrangement is extraordinary and exemplary of the way in which community-based projects can be organized effectively.

Most community-based projects are not nearly as sophisticated in their organization. The scope of project work is typically constrained by meager local budgets and limited expertise. Nonetheless, such projects have an enormous advantage over top-down administration, because they are conceived, administered, and executed from local inertia by local people who take the initiative and dedicate their services to make ecological restoration happen. They realize how much they and their community will benefit from the ecosystem that is rescued from impairment and rewoven as a functional, dynamic entity. Once completed, these same citizens will be proud of their handiwork and will form a political network for its preservation and management. This local commitment to the land is direct and intense and cannot be duplicated by restoration that is bureaucratically conceived and administered by an external agency. Local restoration projects generally satisfy personal and cultural values. For example, volunteers experience the satisfaction of personal response to environmental crisis as described in chapter 2, and the act of participation strengthens social capital in terms of interpersonal bonds and community cohesiveness.

Locally conceived and executed projects have another advantage. They do not necessarily have to be performed in accord with strict budgets and narrow timeframes. Instead, much of the work is conducted by volunteers or by personnel who are engaged by local organizations and paid with locally budgeted funds. Therefore, a project can be performed using the tactics of assisted natural regeneration (chap. 8) that are usually less expensive than technical solutions and may take longer to accomplish. The restoration product from assisted natural regeneration is less intrusive and initially more natural than are projects that rely on technical solutions. Even more important, there is ample time for aftercare, once planning tasks have been implemented. Project sites can be visited frequently by local residents. If the need for midcourse corrections is detected, even long after formal project works has terminated, they can be addressed immediately by local initiative. Such attention may mean the difference between a restored ecosystem that recovers indirectly attained attributes (table 5.1) and one that functions imperfectly and unsustainably, creating tension among all concerned.

The attention to detail that can be devoted in a locally sponsored project will ensure that it provides a broad array of natural goods and services that are essential for the well-being of rural communities in less affluent regions of the world. These natural goods may include wood for construction, thatch, fiber, firewood, and others that were listed in table 2.1. Local projects also provide local natural services, such as the stabilization of mountainous slopes that are otherwise subject to landslides that block roads or bury entire villages.

One of the best known local, bottom-up projects is the North Branch Prairie project described in *Miracle under the Oaks*, by William Stevens (1995), and critiqued by Peter Friederici (2006) and Paddy Woodworth (2013). The project was initiated in 1977 by Steve Packard and a small group of environmental activists near Chicago, Illinois. Packard approached a public official in the Cook County Forest Preserve District and asked whether they could volunteer to clean up trash, cut some brush, scatter some seeds, and generally refurbish degraded prairies on land that the district owned. District personnel wanted to begin such work themselves but were hampered by a lack of funds, and they accepted Packard's offer. The work began and soon attracted other volunteers. The idea of restoring Chicago's former ecosystems spread like a prairie wildfire. Soon, hundreds of citizens were spending their free time working alongside Packard, essentially without plans or administrative structure. By 1993, more than 3,000 volunteers had restored more than 6,700 hectares of degraded prairie and associated oak savanna in an amazing display of civic altruism. We caution, though, that stakeholder issues flared up, as mentioned earlier, which removed some of the luster from this otherwise remarkable effort (Shore 1997).

The use of prison labor for ecological restoration, as mentioned in VFT 1, presents an unusual and potentially beneficial opportunity for locally administered projects. Little can be said definitively, because rigid regulations stymie attempts to determine the effects of restoration work on prisoner populations, and concerns for privacy of individual prisoners must be respected. What appears to be happening, though, is that prisoners become personally involved in restoration project work, and, some of them at least, respond positively when working on socially significant restoration projects. We can't say for sure that such engagement contributes to their rehabilitation, but that possibility exists and presents a superb opportunity for investigation by social scientists. At a minimum, prisoners learn trades and crafts at restoration sites that will assist them with future employment.

## Centrally Administered Projects

Top-down, centrally administered projects are sponsored, planned, and managed by an external agency, institution, or other organization. Local input is negligible and may be limited to the engagement of local contractors. In other words, the local community has no authority and little if any responsibility for the project. The external organization in charge may or may not accommodate local concerns expressed in public hearings or issues raised by local officials. For the most part, centrally administered restoration projects are public works projects that have regional, national, or sometimes international significance. Such projects are conducted, financed, or overseen by provincial or national governments; by transnational organizations; or by NGOs with national or international missions. Distrust between local entities and a central administration sometimes surfaces, as was evidenced in public perceptions that were solicited in connection with VFT 5. Nonetheless, relations can remain cordial and productive, as has been the case so far in VFT 7.

Centrally administered projects are generally larger in scope than a local community can undertake in terms of geographical extent, technical complexity, budgetary needs, and regional coordination. Many are landscape-scale (or larger) restoration programs, consisting of several to many, often contiguous ecological restoration projects. The restoration of a river system, for example, requires central planning and considerable coordination that cannot possibly be accomplished by local effort. In a nation such as China, with its centralized government, limited lands that are suitable for agricultural and other economic production, and its enormous human population, the opportunity for local restoration projects are severely limited (Mann 2011, 192–93).

Centrally administered restoration projects have several distinct advantages over local projects. They can draw upon a large pool of expertise and command large budgets. They can engage the full resources of civil and agronomic engineering. They have access to the latest scientific advancements. They can be coordinated across bureaucratic boundaries and can be molded to comply with legal and political obstacles. These are advantages of which those engaged in locally administered restoration projects can only dream.

In most countries, external restoration projects are usually implemented by large environmental engineering firms whose managers are obligated to shareholders rather than local people. Timeframes and budgets for project implementation are tight, and technical solutions are opted to ensure project tasks can be accomplished quickly and confidently. Project tasks are generally keyed to provide specific ecosystem services. Many such projects easily qualify as ecological engineering but are not always informed by locally derived reference models that are sufficiently detailed to fulfill the ecological attributes of restoration projects outlined in table 5.1. Instead, emphasis is placed on satisfying short-term objectives that can be met rapidly.

These projects only nominally satisfy personal values of ecological restoration (chap. 2) and occasionally destroy cultural values by, for example, moving villages, inundating archeological sites, or despoiling iconic sites. They enhance ecological values but not necessarily to the degree that local projects would, particularly those external projects that are planned with little regard for a local reference model. They will do a superb job of improving socioeconomic conditions and win respect with regard to providing those specific ecosystems services for which they were designed to accomplish, but they will not necessarily provide the full array of natural goods and services. These drawbacks are quibbles relative to the enormous environmental good that these projects accomplish. Our point is that the next generation of such projects could be improved by paying attention to their potentials for accomplishing ecological restoration that emphasizes reestablishment of historic continuity.

One of us (AC) was the restoration practitioner in charge of a centrally administered project to restore forested wetlands in the headwaters of Dogleg Branch, mentioned in chapter 10. A comparison of the Dogleg project and the one in Chicago mentioned earlier, at North Prairie Branch, is a classic study in contrasts. The underlying difference between them was that the former was an elective project, whereas the latter project required layers of government approvals. It took two years of work to satisfy application requirements to obtain the required government permits for the Dogleg restoration work to begin. Permits were eventually issued after the mining company had conducted a four-year pilot project to demonstrate that native trees could be grown and a two-year ecological inventory of local forested wetlands that served as reference sites (Clewell et al. 1982). Professionals who were involved in the project included mining engineers, mine planners, environmental consultants, native nurseries, project managers, heavy equipment contractors, attorneys, highly placed officials in state government, and large administrative support staffs, but no volunteers.

Let's look at these two projects from the perspective of the restoration practitioner. At North Branch Prairie, almost everyone involved was a restoration practitioner. Steve Packard assumed the role of project director, and he and several others assumed the collective role of project manager as well. The Cook County Forest Preserve District was nominally the sponsor, and its personnel provided skeletal administration. For the ecological reference that informed the project, Packard referred to existing ecological literature, a general knowledge of the few remnant patches of prairie and oak savanna, and the species list prepared by an early naturalist. Practitioners developed project plans by consensus throughout the course of the project. Their administrative mode was collegial. They assumed almost total responsibility for all restoration work. The Cook County Forest Preserve District retained basic authority for the project because the project took place on lands under their jurisdiction. District personnel established the bounds for project work to ensure that it was legal and complied with the district's overall mission. The practitioners enjoyed broad flexibility, responsibility, and authority to conduct the project as they saw fit (Packard 1988, 1993).

The North Branch Prairie project was a grassroots, bottom-up endeavor that was not mandated by a public agency. Instead, the Cook County Forest Preserve District benefited from the broad public support of hundreds of citizens who volunteered their free time as restoration practitioners. This was a superb example of people taking collective responsibility for their own concerns in a manner that nicely reflects the four-quadrant model of ecological restoration (chap. 2). Ecological values were fulfilled directly by the restoration. The motivation for many volunteers was the fulfillment of individual values such as reconnecting with nature and responding to environmental crises, as described in chapter 2. Public celebrations at the restored prairie were described by Holland (1994) and provided evidence of the fulfillment of cultural values and the nurturing of social capital. The restored prairies and oak savannas represent natural capital that provides ecosystem services.

In great contrast, the Dogleg Branch project was conducted by only a few restoration practitioners. The mining company was the sponsor, and its mining engineers assumed the other roles of top-down project administration. Detailed project plans were prepared by the company, which incorporated specific conditions that were required by permit from the State of Florida. Stakeholder involvement was limited to formal hearings that were required by law, in which citizens expressed their disillusionment with mining next to their private property. The intent of the project was to repair unavoidable environmental damage and not to fulfill personal and cultural values described in chapter 2. Ecosystem services were implicated in terms of satisfying state regulations concerning water quality, physical reclamation of the premining terrain, and the recovery of vegetation to its immediate, premining state even though some was degraded. Dogleg Branch restoration was narrowly focused as a "flatland" project in terms of the four-quadrant model in chapter 2 and only satisfied few biophysical and socioeconomic values. One benefit consisted of testing several restoration techniques, some for the first time (Clewell and Lea 1990; Clewell et al. 2000), which have since become standard practice in ecological restoration.

## Evolution of Contexts

Ecological restoration has experienced a number of births as a discipline, each one representing a distinctive geographical region, ecosystem type, organizational setting, or industry. However, it was not as if the entire field of ecological restoration was reinvented each time. Instead, the vision and tactics of restoration were repeatedly borrowed and refitted for use in different situations. For example, the phosphate mining industry in Florida and the peat mining industry in northern Europe adopted ecological restoration at about the same time and adapted it to their respective situations in a parallel manner on mined land.

Ecological restoration was first attempted in the Florida phosphate mining industry in 1978, when several restoration techniques for wetland ecosystems were field tested in small pilot plots. The work was encouraged by regulatory personnel who sought innovation for implementing new state rules for mine reclamation by using native species instead of nonnative introductions. Initial results of the plot studies were encouraging (Swanson and Shuey 1980). Regulatory authorities began issuing mining permits that stipulated that wetland restoration be attempted. Several environmental consultants were hired by mining companies to test restoration methods at different project sites, among them the Dogleg Branch restoration site. Shortly thereafter, the Florida Institute of Phosphate Research (FIPR) was organized as a semiautonomous research and development agency for the phosphate mining industry and was funded from state severance taxes on phosphate ore. FIPR-sponsored conferences and research grants attracted academic personnel who provided scientific validation of these exploratory pilot projects and generally advanced mining technology.

By about the year 2000, the strategies and techniques that worked well on former phosphate mines were developed and refined to the point that regulatory agencies could specify detailed prescriptions for restoration that led to reasonably predictable results. Before that time, all restoration work was based on ecological knowledge derived from intelligent tinkering (chap. 8). In this manner, new restoration methods were attempted and older methods refined. Since that time, restoration has been reduced to applying well-tested methods in a uniform manner with a degree of certainty that approached that of civil engineering projects. Project goals, reference models, and historic trajectories were largely replaced by prescriptive permit criteria and performance standards. In short, ecological restoration of the past was preempted by a series of short-term ecological engineering solutions that used native species to recover ecosystem services (soil stability, water quantity and quality) and habitat for officially listed species of plants and animals.

In northern Europe and parts of Canada, a comparable history evolved in the peat mining industry (J. Blankenburg, personal communication, 2006). Environmental concerns led to exploratory ecological restoration that was soon supplemented with ecological studies by university personnel. Strategies and techniques were refined until the process became a uniform ecological engineering exercise that is overseen by government authorities.

We see a trend in these examples from the phosphate mining and peat mining industries. This trend is driven in tandem by government regulators and regulated industries, both of which desire project brevity and uniformity and an empirical basis for determining project compliance with regulatory norms and consequent release from regulatory surveillance. Exploratory ecological restoration develops the basic strategies and techniques, ecological research and development refines them, and ecological engineers apply them thereafter. The process ensures recovery of essential ecosystem services and a degree of recovery of biodiversity. The process is a flatland exercise that addresses limited socioeconomic and ecological values and no personal and cultural values. The process also reflects current global attitudes to maximize economic growth and development, as called for and defended by the paradigm of neoclassical economics, but tempered with mounting uneasiness about the consequences of resultant environmental degradation (Aronson, Blignaut, de Groot, et al. 2010).

Variations on this basic theme exist. For example, bauxite mining was causing much public concern in Western Australia because it was destroying jarrah (*Eucalyptus marginata*) forest. The mining industry recognized that the public outcry could force the closure of their operations, despite weaknesses in the regulatory process. In an outstanding display of corporate responsibility, two of the largest mining companies, Alcoa and Rocla, opted to regulate themselves and develop a highly sophisticated restoration technology that emphasized biodiversity conservation and recognized some cultural values (Nichols and Nichols 2003; Koch 2007). Much attention has been paid to preserving biodiversity, to the point that a state of the art tissue culture laboratory was established by Alcoa to propagate rare plant species with low reproductive potentials (figure 11.2). Both Rocla and Alcoa now achieve more than 70 percent biodiversity reinstatement as part of their standard operating environment (K. Dixon, personal communication). In this instance, industry efforts are identifiable with holistic ecological restoration. The journal *Restoration Ecology* dedicated an entire issue to jarrah forest restoration in a supplement to volume 15, number 4, issued in December 2007. The Western Australian mine site restoration initiative warrants emulation globally.

Some mine reclamation in the Florida phosphate mining industry and elsewhere approaches the successes from Western Australia. However, the impetus for historic continuity was not espoused by the industry as it was in Australia or required by permitting agencies. Reconstruction projects lack attention to detail that would demonstrate an internally motivated ecological commitment. Instead, regulatory oversight stimulated the initiation of outstanding restoration practices, but the regulatory community lacked sufficient statutory authority to require that restoration be conducted to its fullest potential. Excellent work on mine-site resto-

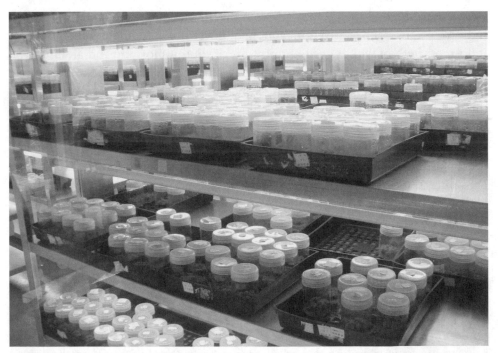

**FIGURE 11.2.** Tissue culture laboratory to grow plantlets of species for which reproduction by seed is difficult or viable seed is scarce at the ALCOA bauxite mine in Western Australia.

ration has been carried out at Richards Bay, South Africa (van Aarde et al. 1998). Increasingly, mining companies operating in tropical forest areas in developing countries, such as Brazil, Panama, Indonesia, and Madagascar are moving closer to restoration as we conceive it in this book. Nautilis Minerals, alone among mining companies preparing to mine minerals and fossil fuel deposits lying 1,000 to 2,500 meters under the surface of the world's oceans, is expressing interest in restoration. Affected marine sea ecosystems will inevitably be deeply scarred by mining. They lie in extraterritorial waters where no legislation applies, and for which no international treaties currently exist. Self-regulation of the industry in this last Wild West is the only option available at the moment.

## Stewardship Models

Many NGOs and some CBOs conduct long-term stewardship programs that consist of ecosystem management that sometimes includes ecological restoration. When that happens, the restoration elements are administered as local initiatives. Volunteers commonly provide much of the labor, including practitioner tasks in restoration projects, as was described for the North Branch Prairie project. Sometimes restoration responsibilities are outsourced, at least in part, to firms that specialize in ecological restoration. Some stewardship programs are sponsored by public agencies whose mission is to protect and manage public lands, such as forest reserves, wildlife refuges, and parks. Financial support for these programs is provided as part of the appropriations of public funds to these agencies.

Common lands under the control of traditional cultures may undergo ecological restoration in the same manner. A tribal council of elders may be the administrative body, which consists of ordinary citizens who are respected for their age or sagacity. Restoration work may be performed by an entire community or tribe under the direction of the elders and as a form of community service. The sacred grove shown in figure 2.3 is being restored largely by eager students from a nearby tribal school. In the United States, restoration is sometimes administered by federal land management agencies and contracted to Native American tribes that use public lands in a traditional manner (Anderson 2005).

## Practitioner Certification

In the title of this book we identified ecological restoration as an emerging profession. A profession commonly meets a number of criteria, five of which are as follows. (1) A profession is identifiable in terms of providing a specific service or product. (2) Those that practice the profession have undergone training or apprenticeship that qualifies them for their professional work. (3) These professionals accumulate experience in their professional practice, which increases their expertise and thus their qualification as practitioners. (4) Professionals form associations to establish thresholds and standards of training or demonstrated ability and experience for entry and promotion within the profession, and to set the standards of quality for professional practice. Associations set other norms for their professions, such as the boundaries that define the limits of that profession. Associations give formal recognition to their members in accordance with their professional credentials. (5) Members of the profession conduct their practice according to a code of ethics that they themselves develop through their professional association. This code requires professionals to meet or exceed the standards and norms set by the professional association in their practice, to keep current with advancements in the knowledge and techniques in their profession, to respect one another and their professional association in matters relating to their profession, and to interact honestly and forthrightly with clients and others with whom they engage professionally. None of these five criteria have been formally developed with regard to ecological restoration. Although it may be premature to identify ecological restoration as a profession just yet, a trend in that direction is pronounced.

Professionals are usually identified because they are either certified by their professional association or licensed by a governmental licensing board. Certification is a declaration by a professional association (or another, generally nongovernmental source) that an individual has attained a particular level of competence as evidenced by completion of training, passage of an examination, or some other measure. Certified practitioners can use professional certification to attract clients or procure employment. Individuals obtain certification by presenting their credentials for review by an independent professional certification board that was established by a professional association. Members of the review board evaluate the credentials of applicants and award certification to those who qualify. Review boards generally function independently of the governance of their parent professional association in order to guarantee impartiality and to protect the privacy of applicants, particularly those who are not members of that association. The board consists of the candidate's peers—usually senior members in the profession. Criteria for certification reflect the norms and standards of the profession that were adopted by the professional association. Periodic recertification may be required to assure that certified individuals keep current in their field and to allow individuals the opportunity to be upgraded in their level of certification as their professional capabilities grow.

Licensing is a governmental function that gives an individual permission to do something that requires a license and is otherwise forbidden. The award of a license is a government privilege and not an individual right. In many, well-established professions, certification is replaced by licensing, whereby a governmental board issues a license that allows the professional to practice within its jurisdiction. Government licensing boards are generally staffed by, or work closely with, senior professionals in the profession. The criteria used for issuing licenses may be developed or sanctioned by a professional association, commonly in concert with academic leadership in that field.

Currently, so far as we know, there is no country where ecological restoration is a formally recognized profession. No governmental body licenses professional restoration practitioners. Certification and licensing is available in some related fields, for which restoration practitioners can apply. For example, a panel of the Australian Association of Bush Regenerators (AABR), confers practitioner status on those who have attained a requisite level of experience in the field. Bush regeneration in Australia is tantamount to ecological restoration and could be considered a subset of it that pertains to recovery of ecosystem types found in Australia. The Society of Wetland Scientists, The Ecological Society of America, and The Wildlife Society are three professional associations that offer certification that some restoration practitioners already hold. The American Society of Landscape Architects (ASLA) participates with state governments in the United States to license practitioners of that profession, and many restoration practitioners are also licensed landscape architects.

A principal intent of professionals and their professional associations is to furnish services, products, or performances of consistently acceptable quality. Once a profession becomes known for the dependable quality of its work, its members will be hired, contracted, or employed and will reap commensurate repute, respect, and compensation. Public agencies commonly insist that certain kinds of work be performed or overseen by a certified or licensed professional. Many private sector firms prefer to engage certified or licensed professionals as employees and contractors. Sometimes these firms must engage these professionals to satisfy conditions stipulated in the provisions of contracts and governmental permits. Government bureaus may insist on engaging certified or licensed professionals. By doing so, these public agencies cannot be faulted for sanctioning what could be perceived as substandard work. For all of these reasons, there are distinct advantages to becoming a certified or licensed professional and to join, support, and participate in the activities of professional associations.

Many public works projects, compensatory mitigation projects, and mine reclamation projects are required by law to be overseen by licensed civil engineers or sometimes by landscape architects or others who are licensed in other professions. Such projects may consist in large part of ecological restoration work. However, licensed professionals must take ultimate responsibility for preparing project plans and their implementation. Public agencies require project site work to be supervised by licensed professionals who oversaw the preparation of restoration project plans that bear their seals (i.e., have impressed pages therein displaying their up-to-date and valid license number). Restoration practitioners are placed at a disadvantage of having to comply with the norms and standards imposed by other professions whose missions are not primarily, if at all, concerned with ecological restoration. Restoration practitioners may make recommendations with regard to project planning and implementation, but ultimately they must comply with the decisions of other professionals.

The certification of professional restoration practitioners would identify those whose competence was recognized and respected by their peers. Potential clients would be more likely to seek certified practitioners for project work. Once a certification program became established, government agencies with responsibilities for managing natural resources could require that restoration work that is performed under their aegis must be conducted by certified professionals. Likewise, government permitting agencies could require that restoration performed for satisfaction of specific permit conditions must be conducted by certified professionals. Government agencies and larger organizations could require that new employees hired as restoration practitioners or planners must be certified. We cannot say how long it will take to get there, but we can be fairly certain that professional certification would improve a practitioner's chances for work or employment in most or all countries with burgeoning markets for their services.

More important for the field of ecological restoration, the very existence of a certification program will be a strong statement that ecological restoration is no longer a diffuse, poorly defined movement or a curious sidelight. Instead, it is a mainstream activity that demands serious consideration and integration in large-scale programs of environmental concern. For that reason, a much broader array of professionals in related fields, as well as government officials, will want to become more knowledgeable about the field. More universities will want to offer courses in it.

There are drawbacks to certification and particularly to licensure. One is that a certification program can be manipulated by unscrupulous individuals who could find ways to discourage application or increase requirements so that only a few elitists would be certified. These individuals could, in turn, monopolize opportunities to secure project work. Certification must be inclusive, not exclusive. Another drawback is that standardization could be encouraged and innovation discouraged, which would be damaging to a profession like ecological restoration that is interdisciplinary and thrives on originality.

A certification program should create a community of certified professionals who intend to improve themselves as well as the restoration profession for the remainder of their careers. Certified professionals should be active participants and advocates of their professions and should closely guard against any intentional or inadvertent actions that would weaken the profession and the certification program. In other words, certification should not be considered as a credential that a restorationist adds to his or her curriculum vitae or resumé. Instead, it should intensify one's dedication toward the restoration profession and should motivate participation in one's professional activities.

The Society for Ecological Restoration (SER) has nearly completed plans for a Practitioner Certification Program (PCP) in ecological restoration, which is scheduled to be launched shortly after this book appears in its second edition, assuming funding is available. We are encouraged by the principles on which the PCP has been drafted and its structure. The PCP is to be administered by a separate corporation, which will operate independently of SER, its parent organization. The leadership of that corporation will be entrusted to restorationists who have had significant experience as practitioners and who themselves are certified.

An important criterion for eligibility for certification in SERPCP is actual project experience as a practitioner. Credit toward experience can be earned in part from serving as a practitioner in related fields, including rehabilitation, ecological engineering, mitigation project work, reclamation, and others. An applicant for certification must be credentialed and knowledgeable. Being credentialed means having completed some type of academic training in one of a number of possible degree programs. Being knowledgeable means that the applicant demonstrates a level of proficiency in particular areas relevant to restoration practice. Such demonstration can consist of university course work, vocational training, in-service training, workshops, or self-study that can be documented. Essential areas of knowledge include biological science with emphasis in plant science, aquatic biology or wildlife biology; ecology; ecological restoration; hydrologic science, soil science; quantitative science and sampling design; planning and project management; and restoration-relevant social sciences. Competence in some of these areas can be demonstrated by completing workshops or in-training courses. The applicant must sign a code of ethics and be knowledgeable about the content of SER's foundation documents, including the *SER Primer* and the *Guidelines for Developing and Managing Ecological Restoration Projects*, both of which are posted on the SER webpage (www.ser.org).

The design of the SERCPC has been particularly difficult to develop, because ecological restoration is not a typical professional field where one aspect builds logically on another in a vertical structure. Ecological restoration is a horizontal field that borrows knowledge and methodology from an array of allied fields. Most practitioners are knowledgeable in some but not all of these fields. Experience is more relevant, whereby it is essential that practitioners have a working knowledge of several fields of knowledge and are adept at integrating these fields in an ecological restoration project. We urge those who may be interested in

SERCPC to watch for information on the SER webpage. Initially, the SERCPC will be available in North America and will be expanded to other parts of the world as soon as possible.

Readers should be advised that the SERCPC has adopted a broader definition of *practitioner* than we have in this book; it encompasses nearly anyone who is involved in restoration project work. All technical and administrative personnel are included, and perhaps some support personnel would also qualify. Personnel from government agencies who maintain regulatory surveillance over ecological restoration projects, are included. This more inclusive definition makes sense for the Practitioner Certification Program that SER intends to launch, because it encourages all parties involved in restoration projects to be fully knowledgeable and conversant in the principles, craft, and science of ecological restoration. We continue to use the term practitioner in a narrower sense in this book, so that we can distinguish the roles within a restoration project.

# Chapter 12

# Moving Restoration Forward—Together

Perceptive readers will have noted a sense of urgency escaping from between the lines in this book. There is a good reason for that. Indeed, we are deeply concerned about the development of our emerging profession. *Ecological restoration* suddenly appeared thirty-odd years ago, almost like a starburst, and it captured the attention of a whole generation of people who had nearly lost hope that we could ever rescue our environmental heritage from rampant global growth and development that materialized from ecologically indefensible and ethically compromised models. Like a starburst, the energy of ecological restoration has radiated in all directions, as different people interpreted it in different ways to address different sets of issues. It has stimulated extended debate among environmental philosophers (see Higgs 2012, Katz 2012, and references therein), which represents a widening ripple that ecological restoration is making in human affairs.

Those who have been drawn to this discipline and this field of action from various directions over the past three decades have neither agreed upon how ecological restoration can be used most effectively nor even as to what its boundaries are. In short, our discipline as yet lacks discipline. A major rationale for writing this book has been to refocus our collective energies before we lose momentum and the starburst dissipates before our eyes.

We begin this chapter by reviewing fundamental perceptions of ecological restoration. Next, we address two topics that have recently served as flashpoints among those who hold divergent perspectives on ecological restoration. These are the relevance of so-called novel ecosystems to restoration and the bearing that ecological restoration brings to climate change issues. The chapter concludes with recommendations that are designed to get us moving together in a direction, and perhaps to a place, where the promise of ecological restoration can be fulfilled.

## Perceptions of Ecological Restoration

An overheard phone conversation:

> "We had a great time, Mom. When I took Helen and the kids into that forest, you should have seen their jaws drop as they looked up at those giant trees." (pause) "Yeah, I got some fishing in. You know, they stocked that river, and the trout were hitting every bait you could throw at them in the plunge pool below the dam." (pause) "No, they didn't. Helen took the kids off to that youth camp nearby. They made hats from reeds they cut by the lake." (pause) "No," (chuckle) "the reeds will keep resprouting. The kids were so proud—when it started raining, the hats they made actually kept their heads dry!"

This fictitious vignette about a family outing illustrates three conflicting viewpoints most of us hold regarding our personal relationship with Nature. These perspectives are not banal, nor inconsequential. They represent the sources of ambiguity regarding what we mean by the term ecological restoration. We identify the three perspectives, or approaches, in that vignette as the Legacy, Utility, and Recovery models of ecological restoration. We are struck by the legacy of unmanaged Nature as represented by giant trees that make our jaws drop. We impound streams and stock them with fish to improve Nature for our utility. We cut reeds in a manner that ensures natural recovery as we rely on Nature to sustain us.

These overlapping perspectives influence us—it goes without saying—as we try to somehow restore impaired ecosystems. We may set out to restore ecosystems in order to recover ecological legacy, to recover flows of ecosystem services, or to recover a mutually sustainable relationship with Nature. It is tempting to say that we do all three when we restore ecosystems, and there would be some truth to it. However, our initial motivation to restore is usually related to only one of these perspectives, not all three. As employees or contractors, the desires of our employers and clients strongly color our motivations. As members of a conservation organization, we adopt our group's mission statement and assume the group's stated priorities as we design restoration projects. If by chance we are associated with a university department, our personal research interests will affect our perception of ecological restoration and the way we try to teach students to think about it all. Our ambiguity with regard to these three models is not limited to ecological restoration. It concerns our fundamental perception of who we are and how we relate to Nature, as encapsulated by the mix of values that were expressed, somewhat whimsically, in the opening vignette. Let's examine these three perspectives more closely.

## Environmental Legacy Model

The environmental legacy model of ecological restoration advocates the return of ecosystems to a preimpairment state in order to recoup some of our dwindling environmental legacy and, with it, our cultural heritage. The intent of this model is to recover authentic prior historical states of ecosystems by reestablishing former expressions of biodiversity and ecosystem dynamics. This was the principle motivation for most practitioners who attempted to restore ecosystems prior to 1980. Those who continue to explore this possibility contend that ecological restoration allows us to "put it back the way it was." This approach, although heroic and widely appealing in its intent, erroneously assumes that Nature is timeless, wilderness is static, and a restored ecosystem can be reconstituted to a prior, predisturbance state. This was a reasonable assumption in the mid-twentieth century when ecological restoration—in the modern sense of the term—began in earnest, and when the majority of ecologists accepted climax theory rather than a worldview of nature-in-flux. It soon became apparent that we did not know enough about most predisturbance ecosystems to restore them with historic fidelity, even if we wanted to, except in their gross features. Eventually, most restorationists realized that ecosystems were dynamic and evolving on account of natural and anthropogenic drivers. Therefore, an ecosystem, even if it could be restored with historic fidelity, would continue to evolve and assume different temporal expressions of biodiversity.

Few restorationists retain this previous perception of ecological restoration, except perhaps at some subliminal level. Those who cling to the legacy notion tend to reside or were educated in cooler parts of the world with relatively stable environments and modest numbers of species, where ecosystem recovery to a preimpairment state seems feasible. Historical fidelity remains the prevailing popular perception of ecological restoration as presented by the news media. The metaphorical strength of the term restoration reinforces a perception of historical fidelity. The *balance of nature* concept, which continues to be perpetuated in schools and on television, reinforces the legacy model.

## Environmental Utility Model

The utilitarian model advocates repairs to ecosystems in order to increase their capacity to provide specific ecosystem goods and services that are valued by people. This model attempts to solve environmental problems and provide relief, without necessarily recovering an ecosystem to a preimpairment state or reestablishing its historic continuity with a promise of long-term sustainability, although both of those outcomes may eventually occur. Advocates of this model strive to "put back function" that was lost. Such project work is not necessarily informed by an ecological reference, except with regard to diminished ecosystem services. These advocates tend to refer to our discipline as "ecosystem" restoration rather than "ecological" restoration. This practice emphasizes the product of restoration in terms of its ability to provide services rather than the process of restoration in terms of the recovery of biodiversity and historic continuity. We concur with Eric Higgs (2012, 82) who expressed this distinction succinctly: "Restoration is a process (not a product) that involves human assistance (not control)." He explained that, "ecological' signals both the focus (ecosystems) and process (integrative)." The environmental utility model seems more closely allied to rehabilitation and to ecological engineering than to ecological restoration. Afforestation and improvements of formerly overstocked natural rangelands for use by domestic livestock are common goals of environmental utility projects. Services targeted for recovery include erosion control, water retention, and improvements in water quality, provision of habitat for officially listed species and other desirable wildlife, and enhancement of recreational opportunities and aesthetic properties of sites such as roadways and former landfills and quarries.

Utilitarian restoration projects fulfill the missions of public agencies that are charged with environmental resource protection and management. Projects can be structured to fit budgets and reach completion within acceptably brief timeframes. In the past, before ecological restoration gained popularity, these same activities were called by other names, such as range management, fisheries management, wildlife management, forestry, civil engineering, and agronomy, among others. The restoration moniker is presently in vogue and is therefore used by many as the designation of choice for such activities. Unfortunately, this broadbrush usage of ecological restoration lacks clarity and hinders cooperation and evaluation. We recommend that utilitarian projects be designated as rehabilitation rather than restoration, if historic continuity is not reestablished. In many cases, where confusion exists, and many ameliorative actions are taking place at the landscape scale, the phrase "restoration *and* rehabilitation" is a very useful one. As we will see below, the term *restoration of natural capital* also helps clear up confusion and promote cooperation.

## Ecological Recovery Model

The ecological recovery model for ecological restoration advocates the reestablishment of an impaired ecosystem's historic continuity in a self-sustainable manner. The motivation for those who advocate the recovery model of ecological restoration is to "put it back on track," which alludes to the analogy of an impaired ecosystem to a derailed train, or to "put it back on course," alluding to relaunching a grounded ship—two images the reader will recall from chapter

4. The ecological recovery model is the one that is presented in the *SER Primer on Ecological Restoration* (SER 2004), but not by name, and the one that we advocate in this book.

The recovery model, like the legacy model, adopts the preimpairment state of an ecosystem as the starting point for planning a restoration project, but, unlike the legacy model, it accepts contemporary conditions and constraints and their effects on the trajectory of a restored ecosystem. The recovery model recovers ecosystem services but in greater array than those that are specifically targeted by a utilitarian project. In addition, the recovery model reestablishes historic continuity as informed by a reference model, and it strives to return an impaired ecosystem to a condition of ecological complexity and self-organization, which leads to increased resilience, dynamic self-sustainability, and biosphere support. In short, the recovery model embraces the realistic goals of the legacy model, provides all of the benefits of the utility model, and returns an ecosystem to wholeness, as defined in chapter 5.

## Novel Ecosystems

New assemblages of species are increasingly common, and they are frequently referred to as "novel ecosystems." These are defined by Hobbs et al. (2009) as ecosystems "in which the species composition and/or function have been completely transformed from the historic system: such a system might be composed almost entirely of species that were not formerly native to the geographic location or that might exhibit different functional properties, or both." These same authors (Hobbs et al. 2009) also proposed the related notion of "hybrid ecosystems" and defined them as those "that retain characteristics of the historic system but whose composition or function now lies outside the historic range of variability." The issue of novel and hybrid ecosystems captivated the attention of many ecologists and science writers. In 2011 alone, over two hundred papers were published in scientific journals with these terms in their titles.

While the term novel ecosystems is new (Hobbs et al. 2006; Lindenmayer et al. 2008), the concept is not. Novel ecosystems is but the most recent in a long list of related concepts that convey essentially the same idea. Among the many terms for this notion, we prefer "emerging ecosystems" (Milton et al. 2003) to novel ecosystems, because emerging more clearly evokes the evolving nature of these assemblages in a neutral fashion, whereas novel, in English, has a connotation of something positive and trendy.

Novelty in ecosystems is not new; it has been occurring since before land plants emerged from the sea. For example, during the Pleistocene, new species combinations arose through independent range shifts of species in response to climate change (Davis 1976; Graham and Lundelius 1984). Those who are impressed by the sudden appearance of novel ecosystems should not lose sight of the wide temporal variation expressed in the development of natural ecosystems. Forested ecosystems, for example, are continually reforming and rearranging themselves as alluvial river channels sweep laterally, back and forth across floodplains.

We concede, though, that we are living in a period that is unprecedented in the Earth's history, called the *Anthropocene era* (Crutzen and Stroemer 2000; Crutzen 2002). This term conveys the seemingly irrepressible domination by people on Earth's environments—the "human footprint." Paul Crutzen, who coined the term, emphasizes the radical changes that took place starting with the industrial revolution, which began circa 1784 with the patenting of a reliable steam engine. William Ruddiman (2003, cf. Ruddiman and Ellis 2009) emphasizes much earlier impact from deforestation and agriculture that has gathered increasing momentum during the past 7,000 years.

The recent surge of interest in novel ecosystems is timely in terms of developing a broadly based sense of urgency to address the problem they pose for natural resource managers. However, we caution that many authors have succumbed to the enormity of the problem and have overlooked or underestimated our capacity to address it. Too many writers and speakers on this topic assume there is no choice but to surrender and accept novel ecosystems as substitutes and the new norm. In this respect, a novel ecosystem would be considered as one whose historic ecological trajectory had been severed and could not be reestablished in any meaningful sense of that term. Replacement of a natural ecosystem with one that is novel occurs relatively abruptly and results in a radical change in biodiversity. In other words, a novel ecosystem, according to many who embrace the novel ecosystem concept, cannot be restored ecologically, and spontaneous regeneration to a prior state is out of the question. A novel ecosystem, therefore, represents the initiation of a new ecological trajectory.

We don't deny the reality of such substitute ecosystems or reject the idea that they are worthy of serious consideration as an option to recover utility on some severely damaged lands such as agriscapes. What we perceive as a serious fallacy, though, is how often some writers overlook the abundance of natural and semicultural areas throughout the world with ecosystems that are relatively intact, and resilient, and which *could* be revitalized and expanded through ecological restoration, if adequate resources were allocated to both the science and application necessary to get the job done. To help pay the costs, these restoration project sites could be integrated into local RNC programs or into larger-scale restoration programs to develop productive, ecologically sustainable landscapes. The alternative is an unthinkably bland and deteriorating world where crippled Nature can no longer sustain human ambition. This is the price of ecological surrender and a Faustian bargain of enormous proportions.

Part of the problem, we contend, is generational. Restoration ecology is a new discipline, and specialists in the field are more familiar with experimental plots studies than with real world ecological restoration projects. They also express cynicism that global economic systems can be made people-friendly. They underappreciate the strides restoration practitioner have made. This is a predictable growing pain in our discipline that will soon dissipate. On the other hand, many ecological restoration projects are not that impressive, because they have yet to attain ecological maturity and most show signs of being underfunded. Others are, unequivocally, poorly designed or executed and are unlikely to lead to lasting results. But better things lie ahead. Ecological restoration and restoration ecology have progressed to the point that they can meld into a potent force. This professional coalition will be supported by a polity that has nowhere else to turn for relief from the impending economic crisis caused by overreliance on a dwindling global resource base.

One critical concern at issue is that of thresholds. Many ecologists in the "novel ecosystems brigade," as some people playfully call advocates of the concept, assume rather hastily, that putative novel ecosystems have crossed ecological thresholds from which there is no return. However, thresholds cannot be verified in the field nearly as easily as they can be diagrammed on paper. No empirical thresholds have been proposed that distinguish historical ecosystems from novel ecosystems, and, to our knowledge, there is no way to ascertain in advance if an alleged novel state is stable or ephemeral. Much of the current fervor regarding novel ecosystems as that term applies to ecological restoration would quickly subside, we argue, if the principle of historical continuity were more widely appreciated and adopted. *This perspective would reveal that many so-called novel ecosystems are really impaired ecosystems that can be recovered by ecological restoration in the holistic sense of the term.*

In addition, we do not support use of the term *hybrid ecosystem*, and not just because ecosystems don't reproduce sexually. So-called hybrid ecosystems are not necessarily irreversibly modified ecosystems or novel ecosystems in the making. Instead, they apparently represent developmental stages in the historic trajectories of ecosystems that are evolving in response to anthropogenically triggered changes in the environment. This is a presumption on our part, because advocates of the notion of hybrid ecosystems have advanced this concept in theoretical terms from their desktops with scarcely any testing or confirmation from examples in Nature. We see no justification for it, and we object to the term hybrid ecosystem because it is a misleading use of jargon. We extend our criticism to proponents of novel ecosystems, who have yet to provide guidance on how to distinguish them from intact natural and semicultural ecosystems.

We are particularly concerned by those who conflate notions of novel and hybrid ecosystems with the theory and practice of ecological restoration. Proponents of an expanded role for novel ecosystems in restoration theory, and applications thereof, seem to reject in good part the approach of selecting ecological references that are based on historical precedents. This is the antithesis of ecological restoration. If novel ecosystems substitute for historical referencing, and are proposed as their own autoreferences, then ecological restoration is relegated to an exercise in ecological engineering.

Some highly modified ecosystems can serve as reference sites for certain stages of ecological restoration projects, not because they are novel but because they are identifiable as stages in historic ecological trajectories. In such instances, ecosystems in developmental transition would help identify anticipated ecological trajectories and thus the probable courses of change. Multiple references that include such sites can be established to help orient this work. As we have mentioned, other novel ecosystems are more nearly identifiable as impaired ecosystems that are themselves candidates for ecological restoration. An impaired ecosystem is a candidate for restoration and not a reference that guides restoration. The species composition of yet other novel ecosystems has been intentionally altered to the point that they would be better considered as designer ecosystems without historical antecedent and of no use in the preparation of reference models. To offer this hodgepodge as the basis for project planning would confusingly expand the boundaries of ecological restoration and needlessly dilute its meaning. A wide range of opinion over the boundaries of ecological restoration as a discipline cripples its capacity to be a game changer in our efforts to help global society regain a semblance of environmental sanity—and security—in the Anthropocene era.

## Climate Change

The novel ecosystem discussion is clearly related to anthropogenic climate change, and to ignore that nexus leads to confusion by those who ponder these complicated issues. We argue for a potentially huge role for ecological restoration at a planetary scale, not necessarily to resolve climate change issues, but at least to be part of the response to worrisome trends, such as global warming as related to the buildup of greenhouse gases in the atmosphere. Here we present the roles of ecological restoration in helping address climate change issues from the perspective of the recovery model. In doing so, we need to realize, as did Ruddiman and his coworkers (see previous section), that significant human impacts on Earth's climate began at the arrival of the Holocene.

Let us look at a recent period, the so-called Little Ice Age, which cooled Europe and North America and caused unusually cool and severe weather in Asia and elsewhere between 1550 and 1850 A.D. Some scholars (e.g., Nevle and Bird 2008; Nevle et al. 2011) argue that the Little Ice Age was triggered, at least in part, when large areas of formerly cultivated land in the American tropics became fallow after rapid and massive demographic collapse caused by diseases that were introduced from European contact since 1492. Photosynthetic activity by rapid forest growth removed enough carbon dioxide ($\sim 5$ ppm), so that the atmosphere retained less heat, and the average world temperature decreased. This loss of heat precipitated the Little Ice Age, which ended with the onset of the industrial revolution, when the combustion of fossil fuels returned carbon dioxide to the atmosphere. This interpretation suggests that increasing photosynthetic activity in a relatively small region of the terrestrial biosphere, centered in the American tropics, had a substantial effect on climate of the entire planet.

Let us now turn this idea around and consider what could be achieved if millions of hectares of formerly forested land in the tropics and elsewhere were restored to native forest—as indeed is being pursued by the UN's Convention for Biological Diversity under Targets 14 and 15 of its 20 Aichi Biodiversity Targets [UNEP/CBD/SBSTTA/REC/XV/2 - Section I, Para 2 (e); http://www.cbd.int/doc/strategic-plan/2011-2020/Aichi-Targets-EN.pdf]. Part of that activity could take place in Europe, where massive deforestation—with consequent climate change impacts—took place over the past thousand years (Kaplan et al. 2009), as well as in North America and elsewhere. Ecological restoration could indeed be a way to significantly increase photosynthetic activity in the biosphere and thereby reduce atmospheric carbon dioxide and ameliorate climates.

The more atmospheric carbon dioxide that can be sequestered in biomass, the cooler the Earth's surface becomes. However, a distinction can and must be made between simple reforestation with monocultures of fast-growing tree species and holistic restoration (Clewell and Aronson 2006). Restoration ecologists have yet to collaborate with climatologists, hydrologists, soil scientists, and other specialists to calculate rates and amounts of carbon sequestration by different kinds of restored ecosystems at different latitudes. Nonetheless, we can safely say that ecological restoration contributes to carbon sequestration in living and detrital biomass, which, in turn, cools the biosphere.

The case for ecological restoration as a contributor to combat anthropogenic global warming does not end with carbon sequestration. Schneider and Kay (1994) argued that transpiration from the biosphere is especially efficient in dissipating heat. The more transpiration, the cooler the biosphere, as has been documented by evidence from aircraft-mounted thermal infrared multispectral scanners, which recorded reflectance values in different ecosystems and land-cover types. Simplification of ecosystems reduces reflectance of heat into space. Retained heat contributes to global warming. The global climatic regime is dependent on species richness and the capacity of large numbers of coexisting species to dissipate energy effectively by transpiration and in much lesser degree by metabolic reactions. Ecological restoration returns complexity to ecosystems and in all likelihood contributes to the cooling of the biosphere. To the extent that emerging ecosystems and designer ecosystems, such as tree plantations tend to be much simpler than natural systems, the argument is quite straightforward that restored forest ecosystems will be much more effective than monospecific tree farms in terms of thermal reflectance, not to mention all the myriad benefits and added values (chap. 2) in terms of biodiversity and other ecosystem services apart from climate change mitigation.

One such added value is the clear role of ecological restoration to reduce or prevent species extinctions related to anthropogenic climate change and other global changes. Conservation biologists rightly fear that losses of species will take place as the world gets warmer, namely those species that are incapable of migrating to cooler regions—at higher latitudes or higher elevations—and becoming incorporated in other ecosystems. Assisted migration of at-risk species into more favorable ecosystems is seen by some as a way to prevent pending extinction. Others view assisted migration as a threat of intentionally introducing species that could become invasive in recipient ecosystems. There is an extensive literature on this debate that we will not review here. At our present state of knowledge, we do not foresee ecological restoration as contributing to a solution to save taxa by moving individuals or populations of at-risk species, and we don't promote such translocations as a restoration practice.

Instead, we suggest a very different role for ecological restoration that enhances the possibility of species being able to survive in situ, without having to migrate or undergo translocation. That role consists of restoring ecosystems that have the capacity to develop all of the emergent properties listed in table 5.1, namely, a full range of ecological processes, ecological complexity, self-organization, resilience to disturbance, and long-term sustainability. If faced with moderate global warming, most at-risk species would have better chances for survival if they lived in complex ecosystems instead of simplified ecosystems. If we are heading toward severe global warming over the next century, as many experts predict, massive extinction is likely no matter what we do; therefore, restoration should be seen as a means to preserve species under conditions of moderate climate change. If a species is threatened by extinction, it will not happen all at once, unless it is narrowly endemic. A species may face extirpation along that edge of its range of geographic distribution where climate change effects are most severe. There is little we can do about the margins; let us focus restoration efforts instead in the heart of those areas.

## Moving Forward

We turn now to a discussion of how ecological restoration could best be performed from a strategic and pragmatic perspective. The first step is the adoption of an improved administrative context for ecological restoration projects that we call CBO (community-based organization) partnerships. Subsequent steps pertain to practitioner training, the formation of regional restoration centers, and launching an e-journal to serve practitioners. Finally, we present the case for conducting ecological restoration within programs for the restoration of natural capital in order to increase the magnitude and impact of the collective restoration effort to human well-being and augment its appreciation by concomitant increase in social capital. Only in this way will societies and cultures invest economically and politically at levels commensurate with the planetary need.

### CBO Partnerships

In chapter 11, we described and contrasted the two extremes of administrative organization for ecological restoration: the locally controlled, bottom-up context and the externally controlled, top-down context. Are there opportunities to combine the advantages of local and external sponsorship of ecological restoration projects? Can their disadvantages be eliminated? We think so. We suggest that whenever possible local sponsorship and administration should be retained. However, external assistance can be offered in equal partnership from external sources to provide whatever resources are lacking locally. Such resources might include funding, planning, and training local people to conduct restoration tasks. These resources could also include expert assistance to inventory the impaired project site, prepare the reference model, provide equipment for site preparation and installation tasks, and assist with monitoring. They could even include legal counsel and administrative support. Not all of these services would be needed for a single project, but most community-based projects would benefit from several of them.

People in local communities (or elsewhere) are not always aware of environmental degradation they may be causing and its consequences to their well-being. They may be equally unaware of measures they could take to become better environmental stewards. They need to be better informed. Except in dire circumstances, people don't appreciate unsolicited advice and demands from outsiders. Therefore, those who would introduce environmental reform need to earn the respect and trust of a community and stakeholders. Thereafter, the external institution or organization needs to urge the local community to assume as much initiative, responsibility, and authority as possible and vest this authority in a motivated CBO. The initial role of an external institution in many cases may be to ask residents and their leadership what natural resource problems they face and to suggest ways to resolve them in a manner that stimulates local initiative. Thereafter, the external authority or NGO can perhaps work in partnership with local stakeholders on a comprehensive restoration program and provide encouragement, capacity-building opportunities, and resources. The wisdom of this approach is evident in most of the virtual field trips, particularly the temperate rainforest restoration in Chile (VFT 7).

This approach has worked successfully in some parts of Africa and India, where governmental offices have been established to stimulate local interest in developing restoration projects and providing assistance to the extent necessary, but not in a top-down manner that suppresses local initiative and enthusiasm. The Australian example mentioned in chapter 11 nicely exemplifies an effective partnership arrangement. This partnership approach that developed around local nonprofit organizations allows a local community to perform larger and more complex projects than could possibly be attempted using only local resources. Being in a large degree local, the project can better satisfy personal and cultural values than a strictly top-down approach and is more likely to fulfill the criteria of the recovery model for ecological restoration.

### Training

There is great need for qualified practitioners who can conduct ecological restoration competently and effectively. There is a similar need for knowledgeable personnel who can assist public and private organizations, institutions, and communities in the conception, planning, performance, and evaluation of ecological restoration projects. They also should be able to communicate with policy makers, the news media, and the general public on matters pertaining to ecological restoration. Competent, knowledgeable restorationists are needed in all regions of the world to nurture the restoration movement and restoration profession. These restorationists should be conversant in the principles of ecological restoration. They should have direct personal experience in various aspects of project conception, planning, execution, and evaluation. Restorationists with these qualifications would be eligible for professional certification in restoration practice. Collectively, they would form a global community or network that would move ecological restoration forward as a profession in order to realize the global promise that our discipline offers. It would also assure that sufficient numbers of competent people are available to take on the jobs that are already becoming available in large numbers in many parts of the world. This is obviously vital to moving forward as well.

There are several diverse paths to becoming knowledgeable in ecological restoration. Degree and certificate programs in universities are an obvious way, and so are courses offered by vocational schools (fig. 12.1). Intensive in-training courses and workshops are available. Some are presented in coordination with professional conferences. Others are given by public agencies for the benefit of their employees and others who care to attend. Self-directed learning and experiential learning are other fully acceptable ways to obtain knowledge; however, those who choose these routes may have more difficulty in demonstrating their accomplishment than those who can produce a diploma or training certificate. Apprenticeship while working with experienced colleagues is an excellent avenue for learning. In chapter 11 we mentioned technical disciplines in which certified professionals should become knowledgeable. This familiarization process can occur quite effectively within an apprenticeship setting.

**FIGURE 12.1.** Field trip of a class of master's candidates in ecorestoration at Dimoria College, Assam, northeastern India. The rifleman accompanied the students for protection in case of encounters with rhinoceros or Bengal tigers. Instructor A. Clewell stands 3rd from left.

## Regional Restoration Centers

Outreach centers to serve strategically located regions necd to be established globally to promote and facilitate ecological restoration. Each center would be readily accessible and serve as a library and clearing house for technical information with particular regard to regional restoration projects. A major function of each center would be to identify and keep track of all ecological restoration projects that take place in its region. Ideally, these centers would form a network that would provide similar services and present information about ecological restoration in a unified manner, such as that presented in the *SER Primer* (SER 2004) and other foundation documents issued by SER on its website (www. ser.org). SER does not have the capacity to open such centers on its own but could endorse them, inform them as to the content they could offer, and maintain a communications network with them.

Regional outreach centers could be established as offices or institutes within universities, larger NGOs, or transnational institutions. These centers could be funded, at least in part, by users from the private sector and others with grant support or institutional resources. Otherwise, the centers would be open to anyone who needed them for a worthy purpose. A center's director and staff would be experienced in restoration theory and practice. Services that an outreach center could provide include the following:

- provide expertise and oversight for community-based restoration projects.
- sponsor and conduct technical workshops and in-training courses on restoration-related topics.
- provide independent reviews of plans for new projects prior to their approval and identify issues and opportunities for consideration by sponsors and project personnel.
- inspect ongoing ecological restoration projects to suggest ways that practitioners could make improvements.
- disseminate knowledge about ecological restoration from experience gained at regional restoration project sites, including successes attained and mistakes to be avoided.
- assist practitioners to prepare case histories of restoration projects for publication.
- monitor selected regional projects following the cessation of project work to determine long-term outcomes.
- assist practitioners to prepare applications for professional certification.
- inform research personnel of opportunities for collaboration with practitioners.
- transmit relevant research findings to practitioners.
- assist public agencies with technical issues and policy development.
- provide educational outreach to the public.
- prepare press releases on regional restoration projects and otherwise inform the news media about restoration issues.
- participate in public events and celebrations that pertain to ecological restoration.

Restoration outreach centers would be institutions where restoration ecologists could learn what research questions would be worthy of investigation, and how answers to those questions would advance restoration practice. Staff at such centers could put practitioners, landowners, and other stakeholders in contact with researchers who could solve important practical problems. When such collaborations occur, we anticipate that the science, craft, and technology of ecological restoration will actually *leap* forward.

## E-journal

The profession of ecological restoration needs a venue for practitioners to publish case histories for essentially every restoration project that is undertaken anywhere. These are necessary to advance the profession of ecological restoration and to demonstrate to policy makers and the environmentally interested public what has been accomplished in our field. This venue, hopefully, would be rigorous enough to attract at least some interest by the academic community and painless enough for practitioners to contribute to it. We suggest a digitally published e-journal, ideally accepting articles in several key languages.

Publication would be instantaneous online. A detailed format would be prepared, so that practitioners would only have to fill in the blanks to prepare a manuscript. This format would ensure that each case history contained information that was comparable from one project to another, with nothing of importance omitted. The finished product would resemble the virtual field trips in this book but would provide more detail, critical monitoring data, and more photos. An editor would review the manuscript to ensure quality and readability. Some form of peer review may be needed, but it should not be as onerous and intimidating as the peer review process now in place for most scientific journals. Instead, clarification and completeness would be the primary concerns of peer review. Otherwise, few practitioners would likely bother to prepare manuscripts. If regional restoration outreach centers were available, their staffs could encourage practitioners to prepare manuscripts and assist in their preparation. Publications of case histories could be a criterion that leads to practitioner certification or recertification, and a certification program could stimulate the preparation of case histories.

## RNC-Ecological Restoration Linkage

We foresee a great opportunity for a closer link between ecological restoration and the larger arena called the restoration of natural capital (RNC). Stocks of natural capital are already too low to sustain the well-being of people on Earth at present levels of consumption and transformation (Sanderson et al. 2002; Wackernagel et al. 2002). Consequences of insufficient natural capital include hunger, deteriorating public health, political unrest, social chaos, warfare, and genocide. The restoration of ecosystems and ecological landscapes in an RNC context becomes an obvious and essential response and indeed part of the remedy for this predicament and trend (Clewell 2000b; Clewell and Aronson 2006). Existing large regional restoration programs (e.g., those described by Doyle and Drew 2008) tend to be limited in vision, with most emphasis being given to increasing flows of a limited array of ecosystem services. The rationale for restoring for maximum ecological effect by way of the recovery model should be obvious to most everyone knowledgeable with natural resource issues, at least for those who don't wear the blinders of annual budgeting.

Demographic increase and growing rates of per capita consumption promise to worsen the situation, which makes the need for restoration, particularly in an RNC context, even more imperative. Comprehensive studies on the ecological footprint made by humans (Wackernagel and Rees 1996; MA 2005) compellingly show the increasingly essential need for both nature conservation (de Groot 1992; Balmford et al. 2002) and ecological restoration of degraded ecosystems for our well-being and perhaps our very survival. The weight of this argument and the attention paid to it are growing rapidly, as we have summarized elsewhere (Aronson et al. 2007a) and as is portrayed in articles written for broader audiences by Paddy Woodworth (2006a, 2006b; 2013). We call for affluent nations and transnational organizations to strengthen their commitment and expand their vision. Programs should be designed, funded, endowed, and executed to give maximum benefit in terms of ecological recovery and augmentation of social capital. They should be seen as investments in ecological infrastructure and social and human capital, as well as natural capital. James Blignaut (2008) called this "the need for an integrated approach to economic development and restoration." It is time to scale up restoration activities worldwide, as part of the campaigns to preserve biological diversity, combat desertification and degradation worldwide, and to combat anthropogenic climate change—the three great topics inspired by the Earth Summit held in Rio de Janeiro twenty years ago.

* * * * *

Please recall figure 2.6, where the holons in the four-quadrant model of ecological restoration are arching toward a common point. Those holons approach each other as people begin to understand that Nature sustains us and that we must sustain Nature. We can no longer afford to remain separated by greed and bellicose competition. Ecological restoration, whether or not it is performed in concert with an RNC program, provides benefits in an intergenerational manner when we serve as good stewards of our planet. Ecological restoration is our best hope to preserve biodiversity and healthy ecosystems.

This has all been said before, among others, by John Cairns Jr., to whom we have dedicated this book. But his message probably needs a fair number of repetitions with variations on the theme. The time may be getting better for acceptance of his message. We are encouraged in this regard by *The Future We Want*, a document that summarizes the outcome of the Rio + 20 United Nations Conference on Sustainable Development in June, 2012 (www.uncsd2012.org/thefuturewewant.html). This summary statement strongly endorsed restoration and called for "holistic and integrated approaches to sustainable development which will guide humanity to live in harmony with nature and lead to efforts to restore the health and integrity of the Earth's ecosystem." Perhaps people are indeed becoming more amenable to receiving John Cairn's message. With that prospect in mind, we have tried to reach a readership in this book that is now ready to move forward—together!

The definitions below pertain to the usage of terms in this book. Some terms may have broader or additional meanings that are not given. As a courtesy to readers, glossary terms are italicized the first time they appear in the book.

\* \* \* \* \*

**abiotic**. Pertaining to the nonliving or physical aspects of an ecosystem such as soil, moisture, nutrients, and climatic factors.

**agroforest**. Forest or woodland that was intentionally planted, at least in part, to species that provide food or other economic products to people.

**alternative state**. A semicultural ecosystem (usually) that differs from the natural (nonanthropogenic) ecosystem that previously occupied a given site.

**anoxia**. Lacking oxygen, as in saturated soil.

**Anthropocene era**. The current geological period, characterized by the massive human impact on the biosphere. Thought by some to have begun ~ 200–250 years ago and by others 7,000 years ago.

**anthropogenic**. Having an origin that was influenced or shaped by people.

**biodiversity**. The diversity of life at various levels of organization (genetic, individual, population, community, ecosystem, bioregion, biosphere) and taxonomic rank (species, genus, family, etc.).

**basin of attraction.** A term borrowed from field theory in physics and embryology that ecologists have applied to indicate the tendency of species and populations to assemble in particular groupings to form distinctive, predictable communities.

**biome.** A large, regional ecological unit, usually defined by its dominant vegetative pattern, such as the coniferous forest biome of northern Europe.

**biophysical.** Pertaining collectively to organisms and their supporting physical environment, that is, the biota and the abiotic components of an ecosystem.

**biota.** All of the different kinds of living organisms (plants, animals, microbial forms) in a given location.

**certification.** Formal recognition by one's peers of a person's professional capabilities.

**climax.** A mature community or ecosystem that is apparently stable under existing environmental conditions and therefore persistent in its present state. Sometimes called *old-growth* by foresters to avoid identification with the largely discredited climax theory.

**coevolution.** Mutual exertions of selective pressures by two or more species during the course of evolution that imbue each species with adaptive traits when these species occur together in the same community.

**community structure** (or simply *structure*). The physical appearance of a community as determined by the sizes, life forms, abundance, and distribution of the predominant plant species.

**community.** The biota of an ecosystem or a particular portion of it, such as the plant community, insect community, or epiphytic community.

**creation.** The intentional replacement of an ecosystem with another kind of ecosystem of alleged greater value, as has been required commonly for satisfying compensatory mitigation requirements.

**cultural ecosystems.** Ecosystems that have developed under the joint influence of natural processes and human-imposed organization.

**degradation.** The incremental and progressive impairment of an ecosystem on account of continuing stress events or punctuated minor disturbances that occur with such frequency that natural recovery does not have time to occur.

**desertification.** Degradation that causes a site to become progressively drier, relative to its undegraded state, although true desert conditions are not necessarily attained.

**designer ecosystems.** Intentionally created biotic assemblages whose species have been selected in the design process to serve a specific purpose.

**disturbance.** A natural or anthropogenic event that changes the structure, content and/or function of an ecosystem, usually in a substantial manner (also called *perturbation*). Alternatively, one incremental event in a sequence that causes ecosystem degradation.

**ecological attributes.** Biophysical (composition, structure, abiotic/landscape support) and emergent (functionality, complexity, self-organization, resilience, self-sustainability, biosphere support) properties of ecosystems.

**ecological engineering.** The manipulation and use of living organisms or other materials of biological origin to solve problems that affect people.

**ecological footprint**. Any measure of human demand on the biosphere to provide resources and absorb wastes.

**ecological restoration**. The process of assisting the recovery of an impaired eco-system.

**ecology**. The study of interactions among living organisms and between organisms and their environment.

**ecophysiology**. The study of how an organism responds physiologically to environmental conditions.

**ecosystem**. The complex of living organisms and the abiotic environment with which they interact at a specified location.

**ecosystem health**. State or condition of an ecosystem in which its dynamic attributes are expressed within "normal" ranges of activity, relative to its ecological stage of development.

**ecosystem management**. The manipulation of natural areas by technical managers to maintain ecosystem integrity and health.

**ecosystem services**. The benefits of nature in support of people, households, communities, and economies.

**ecotone**. A transition zone between ecosystems.

**flatlander**. Said of persons with a narrow and nonholistic approach to ecological restoration, specifically with interests in only one quadrant (or only in certain elements in both of the objective quadrants) of the four-quadrant model of ecological restoration.

**forb**. Herb that is not a grass or grass-like.

**four-quadrant model**. Model that proposes that ecological restoration satisfies values concurrently of an ecological, socioeconomic, personal, and cultural nature.

**fragmentation**. The division of a formerly continuous natural landscape into smaller natural units that are isolated by intervening lands that were converted for economic production or development.

**fuel load**. The potentially combustible materials, both living and nonliving, in an ecosystem.

**fuelwood**. Wood gathered for domestic heating and cooking.

**function**. Said of the dynamic aspects of ecosystems, such as photosynthesis, primary production, sequestering and recycling of mineral nutrients, and maintenance of food webs. Sometime restricted in meaning to these metabolic activities, and sometimes expanded to include all ecosystem processes.

**functional group**. Two or more species in the same ecosystem that carry out the same function or respond in similar ways to a given stress.

**Gaia**. Concept that considers planet Earth and all of its living beings to be interconnected as a single, self-regulating organism, or as a single, self-organizing whole. The concept can be considered as real, allegorical, or metaphysical.

**garrigue**. Species-rich, pyrogenic, and anthropogenic shrublands of Mediterranean climates that occupy a variety of soils around the Mediterranean Basin of Europe, northern Africa, and Asia Minor.

**geomorphology**. The description and study of land forms.

**herbivore**. An animal that feeds on plants. *Herbivory* is the state or condition of feeding on plants.

**historic fidelity**. Resembling an ecosystem as it appeared in the past.

**historic continuity.** Said of a restored ecosystem that resumed its development following impairment as a continuation of its former ecological trajectory.

**holistic ecological restoration**. Recovery of an impaired ecosystem to ecological wholeness by reestablishing historic continuity and recovering ecological attributes, thereby fulfilling concomitant human values.

**Holocene era**. The geological period ranging from the end of the last major glaciation, ca. 11,700 years ago, until the present day.

**human well-being**. The state whereby a person experiences general contentment with living conditions and is able to pursue modest goals with a reasonable likelihood of success. Wealth is not necessarily a precondition of well-being.

**hydrology**. The study of hydrodynamics, including the input, retention, output, and recycling of water.

**hydroperiod**. The duration that a soil or substrate is saturated or inundated over the course of a year or other time period.

**hypha**. (plural **hyphae**). A long, fine, branching filamentous structure of a fungus, commonly single celled and multinucleate.

**impact**. A disturbance or other harmful occurrence to an ecosystem or landscape, which is caused by intentional or inadvertent human activity.

**impairment**. The state or condition of an ecosystem or landscape that has been degraded, damaged, or destroyed as a result of extraordinary impact or disturbance from which spontaneous recovery to its former state is unlikely, at least in the short term.

**indigenous**. Native to a given location.

**intact**. Said of an ecosystem that is whole and not impaired.

**invasive species**. A nonnative species (usually) whose populations proliferate at the expense of native species and co-opt space and habitat that would otherwise be occupied or recaptured by native species. If native, the species occupy an unusual landscape position on account of prior impacts.

**keystone species**. A species that has a substantially greater positive influence on other species than would be predicted by its abundance or size.

**K-strategist**. Technical term used in ecology referring to persistent, generally long-lived plants that are either competition tolerant or—in high-stress environment—stress tolerant and that dedicate their energy reserves to the formation of vegetative rather than to reproductive structures. Cf. *r-strategist*.

**landscape**. An assemblage of ecosystems that are arranged in recognizable patterns and that exchange organisms and materials such as water.

**life form**. The distinguishing features of a plant, such as woody or herbaceous, evergreen or deciduous, spiny or spineless.

**local ecological knowledge (LEK)**. Current and ever-expanding, useful knowledge about species and ecosystems, as gathered by people who live in rural landscapes in a sustainable manner. See also *TEK*.

**manipulation**. A direct intervention at a project site by a practitioner.

**mesic**. Said of a terrestrial ecosystem with soils that are generally moist, rather than dry (xeric) or wet (hydric), or of a species that occurs in a mesic habitat.

**microclimate**. Ameliorated atmospheric conditions, relative to those of the macroclimate in the region, caused by community structure (shade, windbreaks, etc.) and processes (e.g., transpiration) in an ecosystem.

**mitigation**. An approach or strategy used by government agencies to require that unavoidable environmental damage is compensated by ecological restoration or another activity (rehabilitation, reclamation, enhancement, etc.).

**monitor**. Systematically gather information on an ecosystem, sometimes repetitiously and using a standard protocol, in order to determine the degree of attainment of performance standards or goals.

**mycorrhiza**. A mutual association between a plant root and a fungus in the soil whose hyphae (strands) penetrate the root and extract carbohydrate while providing the root with phosphorus and other mineral nutrients.

**natural capital**. Stocks of natural resources that are renewable (ecosystems, organisms), nonrenewable (petroleum, coal, minerals, etc.), replenishable (the atmosphere, potable water, fertile soils), and cultivated (crops, forest plantations, etc.), and from which flow natural goods and services.

**natural goods and services** (or **ecosystem services**). Foods, fuels, or other products of economic or cultural value that are supplied by ecosystems, and various economically valuable services that ecosystems provide to people, such as flood-water retention and erosion control—all without costs of production and maintenance.

**nongovernmental organization (NGO)**. A private, nonprofit organization that usually receives funding from philanthropic sources or government grants and that provides services that are otherwise generally unavailable.

**nutrients**. Mineral elements required for plant and microbial metabolism and growth, such as phosphorus, calcium, magnesium, iron, etc.

**objective**. A specific, short-term, and direct result that is desired from project work, which will contribute eventually toward the achievement of project goals.

**outplant.** To remove growing stocks of plants (e.g., seedlings) from a plant nursery and relocate them to a project site.

**patch dynamics.** A conceptual approach to ecosystem and community analysis that emphasizes dynamics of heterogeneity within a system.

**performance standard.** A value or threshold condition that is determined by monitoring and that, when attained, verifies that a particular objective has been achieved.

**phenology.** The seasonality of plant processes, such as breaking dormancy, flowering, seed dispersal, and leaf fall as related to time periods on a calendar.

**plankton.** Commonly microscopic algae and animals that lives suspended in a water column.

**practitioner.** Someone who applies practical skills and knowledge to complete restoration tasks at project sites.

**prairie.** The usual term to designate grasslands in North America.

**process.** Dynamic aspect of an ecosystem or landscape, sometimes considered synonymous with *function*, including interactions such as transpiration, competition, parasitism, animal-mediated pollination and seed dispersal, mycorrhizal relationships, and other symbiotic relationships.

**production system.** Land or landscape unit, or wetland or marine area, allocated to the production of food, fiber, pasturage, aquaculture, and other marketable commodities that could also be consumed for subsistence, and that is usually maintained with external inputs of energy (e.g., fossil fuels) and materials (e.g., lime, agrichemicals).

**propagule.** Any plant reproductive structure, sexual and vegetative, such as a seed, spore, or rootstock that proliferates.

**provenance.** The geographic place of origin or source of seeds, nursery stock, and other propagules and organisms that arrive at, or are intentionally introduced at, a project site.

**pyrogenic.** Said of an ecosystem that originates by agency of fire and that is maintained by periodic fires.

**reallocation.** The rededication of ecosystems for new uses of an economic type, other than the transformation of an ecosystem to an alternative state.

**reclamation.** Conversion of land perceived as being relatively useless to a productive condition, commonly for agriculture and silviculture. Recovery of productivity is the main goal.

**reference.** One or more actual ecosystems (called reference sites), their written ecological descriptions, and/or information from secondary sources (e.g, historical photographs or accounts, paleoecological data) that serve as a basis for guiding the development of an ecological restoration project.

**reference model.** An ecological description of an ecosystem to serve as a basis for preparing restoration plans, which was derived from the study of reference sites and/or from secondary sources of information.

**rehabilitation.** The recovery of ecosystem processes to regain normal function and ecosystem services without necessarily restoring the biodiversity of the reference or its projected trajectory.

**resilience.** The capacity of an ecosystem to tolerate or fully recover spontaneously from disturbance.

**restoration ecology.** The science upon which the practice of ecological restoration is based and that provides the concepts and models on which practitioners depend. Alternatively, the science that advances the frontiers of theoretical ecology through studies of restored ecosystems and those that are undergoing restoration.

**restoring natural capital (RNC).** Investment in natural capital stocks and their maintenance in ways that will improve the functions of both natural and human-managed ecosystems, while contributing to the socioeconomic well-being of people, through holistic restoration of ecosystems, ecologically sound improvements to lands managed as production systems for useful purposes, improvements in the utilization of biological resources, and the establishment or enhancement of socioeconomic systems that facilitate the incorporation of knowledge and awareness of the value of natural capital into daily activities.

**revegetation.** Establishment of plant cover on open land, usually with one or few species, irrespective of their provenance.

**rhizome.** Stem of a plant that grows underground or within a substrate under water Also called a rootstock, although it differs anatomically from true roots.

**riparian.** Pertaining to rivers, such as a forest that occupies a river floodplain.

**r-strategist.** Technical term used in ecology referring to plants that are short-lived; weedy or opportunistic; that colonize open and disturbed environments; that are intolerant of competition; and that expend their energy reserves on reproduction rather than on vegetative development. Cf. *K-strategist*.

**ruderal.** Weedy.

**runoff.** Rainfall or other water that moves toward lower elevations by spreading across the land surface, rather than flowing within a defined channel.

**salinization.** Process by which soil comprising the root zone becomes increasing more saline (salty) on account of the evaporation of irrigation water or another cause generally related to land use.

**savanna.** Vegetation consisting of dense grasses or sedges, commonly with forbs intermixed, which is interrupted at wide intervals by shrubs and trees that grow individually or in small clumps or patches.

**sedge.** Grasslike plant that belongs to the family Cyperaceae. Sedges are more common than grasses in many wetlands.

**seed bank** (or *propagule* bank). Seeds (and other propagules) in the soil that can replenish the vegetation following disturbance.

**self-organizing**. Said of an ecosystem that develops and functions in response to its internal processes. Synonym: *autogenic*.

**self-sustainable**. Said of a self-organizing ecosystem that persists indefinitely, although not without transformations in response to its own internal dynamics, to environmental flux, and to longer-term change in environmental conditions.

**semicultural ecosystem**. A natural ecosystem that has been managed and partially altered by human land-use activities in a manner that conserves most elements of biodiversity and ecosystem functionality in a more or less sustainable manner.

**sere**. All developmental stages collectively of an ecosystem as a new ecosystem matures or a disturbed ecosystem recovers. Each stage can be called a seral stage or seral community.

**silviculture**. The establishment and maintenance of trees or a forest, generally for the production of wood or another marketable commodity.

**spatial**. Pertaining to size, dimensions, or location.

**species composition**. All of the different kinds of species that occur at a location.

**stakeholder**. Any person who is affected in any way—positively or negatively, and directly or indirectly—by an activity, including an ecological restoration project.

**state**. The appearance, expression, or manifestation of an ecosystem or landscape as determined by species composition, the life forms, sizes, and abundance of individuals, and community structure.

**stochastic**. Happening by chance.

**stress**. A normally occurring condition or recurring event that is more detrimental to some species than to others, and that largely determines species composition and abundance in an ecosystem. Examples of stress include freezing temperatures, drought, salinity, fire, and unavailability of nutrients.

**subsistence**. The provision of food, fuel, and other essentials for use by an individual, family, or tribal village, as opposed to marketable commodities that are sold or traded.

**succession**. The sequence of stages that occur in species composition (particularly) and in species abundance, community structure, and the complexity of interspecific interactions as an ecosystem develops or as it recovers from disturbance. Cf. *sere*.

**symbiont**. One of two organisms that live in close contact for mutual benefit.

**target**. The intended long-term outcome (endpoint or goal) of a restoration project, which sometimes is not fully achieved until long after restoration project work has ceased.

**taxon.** (plural, **taxa**) A hierarchical category of organisms, e.g., subspecies, species, genus, family, in a system of classification.

**temporal.** Pertaining to time and duration.

**traditional cultural practices.** The application of traditional ecological knowledge that leads to the development and maintenance of cultural ecosystems.

**traditional ecological knowledge (TEK).** Ecological knowledge derived through societal experiences and perceptions that are accumulated within a traditional society through interaction with nature and natural resources. TEK commonly originates through trial and error and is frequently passed down to subsequent generations by oral tradition. See also *LEK*.

**trajectory.** The sequence of biotic expressions of an individual ecosystem in the past.

**vascular plant.** Plants containing vascular tissue (xylem, phloem), including all trees, flowering plants, and ferns, and excluding algae, fungi, lichens, and mosses.

**zooplankton.** Plankton consisting of animal species.

# REFERENCES CITED

Anderson, M. K. 2005. *Tending the Wild Native: American Knowledge and the Management of California's Natural Resources*. Berkeley: University of California Press.

Anderson C., J. Celis-Diez, B. Bond, G. Martínez Pastur, C. Little, J. Armesto, C. Ghersa et al. 2011. "Progress in Creating a Joint Research Agenda that Allows Networked Long-Term Socio-Ecological Research in Southern South America—Addressing Crucial Technological and Human Capacity Gaps Limiting its Application in Chile and Argentina." doi: 10.1111/j.1442-9993.2011.02322.x.

Apostol, D. and M. Sinclair, eds. 2006. *Restoring the Pacific Northwest: The Art and Science of Ecological Restoration in Cascadia*. Washington, DC: Island Press. Aronson, J., P. H. S. Brancalion, G. Durigan, R. R. Rodrigues, V. L. Engel, M. Tabarelli, et al. 2011. "What role should government regulation play in ecological restoration: Ongoing debate in São Paulo State, Brazil." *Restoration Ecology* 19:690-695.

Aronson, J., J. N. Blignaut, R. de Groot, A. Clewell, P. P. Lowry II, P. Woodworth, D. Renison et al. 2010. "The Road to Sustainability Must Bridge Three Great Divides." *Annals of the New York Academy of Sciences* (Special issue, *Ecological Economics Reviews*) 1185:225–36.

Aronson, J., J. N. Blignaut, S. J. Milton, D. Le Maitre, K. J. Esler, A. Limouzin, et al. 2010. "Are socioeconomic benefits of restoration adequately quantified? A metaanalysis of recent papers (2000–2008) in Restoration Ecology and 12 other scientific journals." *Restoration Ecology* 18:143–154.

Aronson, J., F. Claeys, V. Westerberg, P. Picon, G. Bernard, J.-M. Bocognano, and R. de Groot. 2012. "Steps towards Sustainability and Tools for Restoring Natural Capital: Etang de Berre (Southern France) Case Study." In M. Weinstein and E. Turner, eds., *Sustainability Science: Balancing Ecology and Economy*. New York: Springer, 113–40.

Aronson, J., A. F. Clewell, J. N. Blignaut, and S. J. Milton. 2006. "Ecological Restoration: A New Frontier for Conservation and Economics." *Journal for Nature Conservation* 14:135–39.

Aronson, J., S. Dhillion, and E. Le Floc'h. 1995. "On the Need to Select an Ecosystem of Reference, However Imperfect: A Reply to Pickett and Parker." *Restoration Ecology* 3:1–3.

Aronson, J., C. Floret, E. Le Floc'h, C. Ovalle, and R. Pontanier. 1993a. "Restoration and Rehabilitation of Degraded Ecosystems in Arid Land Semi-arid Lands. 1. A View from the South." *Restoration Ecology* 1:8–17.

———. 1993b. "Restoration and Rehabilitation of Degraded Ecosystems. 2. Case Studies in Chile, Tunisia and Cameroon." *Restoration Ecology* 1:168–87.

Aronson, J., and E. Le Floc'h. 1996a. "Que faire de tant de notions du paysage?" *Natures, Sciences, Sociétés* 4:264–66.

———. 1996b. Vital landscape attributes: missing tools for restoration ecology. *Restoration Ecology* 4: 377-387.

Aronson, J., S. J. Milton, and J. N. Blignaut, eds. 2007a. *Restoring Natural Capital: Science, Business and Practice*. Washington, DC: Island Press.

———. 2007b. "Restoring Natural Capital: Definitions and Rationale." In J. Aronson, S. J. Milton, and J. Blignaut, eds., *Restoring Natural Capital: Science, Business and Practice*. Washington, DC: Island Press, 1–2.

Aronson, J., S. J. Milton, J. N. Blignaut, and A. F. Clewell. 2006. "Conservation Science as if People Mattered." *Journal for Nature Conservation* 14:260–63.

Aronson, J., C. Murcia, D. Simberloff, G. Kattan, K. Dixon, and D. Moreno-Mateos. In review. "Are Novel Ecosystems a Slippery Slope for Restoration Ecology?" *Plant and Soil*.

Aronson, J and J. Van Andel. 2012. "Restoration ecology and the path to sustainability." Pages 293-304 in: Van Andel, J. and J. Aronson, eds. *Restoration Ecology: The New Frontier. 2nd edition*. Oxford, UK: Wiley-Blackwell.

Bainbridge, D. 2007. A *Guide for Desert and Dryland Restoration*. Washington, DC: Island Press.

Baird K., and J. Rieger. 1989. "A Restoration Design for Least Bell's Vireo Habitat in San Diego County." In D. L. Abell, ed., *Proceedings of the California Riparian Systems Conference: Protection, Management, and Restoration for the 1990's*. General Technical Report PSW-110. Berkeley CA. Pacific Southwest Forest and Range Experiment Station, Forest Service, US Dept of Agriculture, 462–67.

Bakker, J. P., and T. Piersma. 2006. "Restoration of Intertidal Flats and Tidal Salt Marshes." In J. van Andel and J. Aronson, eds., *Restoration Ecology: The New Frontier*. Oxford: UK: Blackwell, 174–92.

Balaguer, L., R. Arroyo-Garcia, P. Jimenez, M. D., Jimenez, L. Villagas, I. Cordero, E. Manrique et al. 2011. "Forest Restoration in the Lomas (Fog Oases) of Coastal Peru: Genetic and Experimental Evidence Indicate That Cultural Components Should Be Part of the Reference System." *PLoS ONE*. 6:e23004. doi:10.1371/journal.pone.0023004.

Balée, W. 2000. "Elevating the Amazonian Landscape." *Forum for Applied Research and Public Policy* 15:28–33.

Balmford, A., A. Bruner, P. Cooper, R. Costanza, S. Farber, R. E. Green, M. Jenkins et al.

2002. "Economic Reasons for Conserving Wild Nature." *Science* 297:950–53.

Bastow Wilson, J., I. Ullmann, and P. Bannister. 1996. Do Species Assemblages Ever Recur? *Journal of Ecology* 84: 471-474.

Bernhardt, E. S., M. A. Palmer, J. D. Allan, G. Alexander, K. Barnas, S. Brooks, J. Carr et al. 2005. "Synthesizing U.S. River Restoration Efforts." *Science* 308:636–37.

Blignaut, J. N. 2008. "Fixing Both the Symptoms and the Causes of Degradation: The Need for an Integrated Approach to Economic Development and Restoration." *Journal of Arid Environments* 73:696–98.

Blignaut, J., J. Aronson, M. Mander, and C. Marais. 2008. "Investing in Natural Capital and Economic Development: South Africa's Drakensberg Mountains." *Ecological Restoration.* 26:143–50.

Blignaut, J. N., J. van Ierland, T. Xivuri, R. van Aarde, and J. Aronson. 2011. "The ARISE Project in South Africa." In D. Egan, J. Abrams, and E. Hjerpe, eds., *Exploring the Social Dimensions of Ecological Restoration.* Washington, DC: Island Press, 207–19.

Blondel J., J. Aronson, J. -Y. Bodiou, and G. Bœuf. 2010. *The Mediterranean Basin—Biological Diversity in Space and Time.* Oxford, UK: Oxford University Press.

Bonnicksen, T. M. 1988. "Restoration Ecology: Philosophy, Goals, and Ethics." *Environmental Professional* 10:25–35.

Bormann, F. H., and G. E. Likens. 1979. *Pattern and Process in a Forested Ecosystem.* New York: Springer.

Bowman, D. M. J. S. 1998. "The Impact of Aboriginal Landscape Burning on the Australian Biota." *New Phytologist* 140:385–410.

Boyer, K. E., and J. B. Zedler. 1999. "Nitrogen Addition Could Shift Plant Community Composition in a Restored California Salt Marsh." *Restoration Ecology* 7:74–85.

Bradley, J. 1971. *Bush Regeneration.* Sydney, AU: Mosman Parklands and Ashton Park Association.

Bradshaw, A. D. 1987. "Restoration: An Acid Test for Ecology." In W. R. Jordan III, M. E. Gilpin and J. D. Aber, eds., *Restoration Ecology a Synthetic Approach to Ecological Research.* Cambridge, UK: Cambridge University Press, 23–29.

———. 1996. "Underlying Principles of Restoration." *Canadian Journal of Fisheries and Aquatic Science* 53 (Supplement 1): 3–9.

Brancalion, P. H. S., R. A. G. Viani, J. Aronson, and R. R. Rodrigues. In review. "Tree Nurseries and Seed Collecting in Service of Tropical Forest Restoration: How to Increase Their Biodiversity and Potential for Social Integration?" *Restoration Ecology.*

Brewer, J. S. 2001. "Current and Presettlement Tree Species Composition of Some Upland Forests in Northern Mississippi." *Journal of the Torrey Botanical Society* 128:332–49.

Brewer, S., and T. Menzel. 2009. "A Method for Evaluating Outcomes of Restoration When No Reference Sites Exist." *Restoration Ecology* 17:4–11.

Buisson E., and T. Dutoit. 2006. "Creation of the Natural Reserve of La Crau: Implications for the Creation and Management of Protected Areas." *Journal of Environmental Management* 80:318–26.

Burkhart, A. 1976. "Monograph of the Genus *Prosopis*." *Journal of the Arnold Arboretum* 57:219–49, 450–525.

Cabin R. J. 2011. *Intelligent Tinkering: Bridging the Gap between Science and Practice.* Washington, DC: Island Press.

Cabin, R., A. Clewell, M. Ingram, T. McDonald, and V. Temperton. 2010. "Bridging Restoration Science and Practice: Results and Analysis of a Survey from the 2009 Society for Ecological Restoration International Meeting." *Restoration Ecology* 18:783–88.

Cairns, J., Jr. 1993. "Ecological Restoration: Replenishing Our National and Global Ecological Capital." In D. Saunders, R. Hobbs, and P. Ehrlich, eds., *Nature Conservation 3: Reconstruction of Fragmented Ecosystems.* Chipping Norton: Surrey Beatty, 193–208.

Callenbach, E. 1975. *Ecotopia.* Berkeley: Banyan Tree Books.

Clark, W. C., and N. M. Dickson. 2003. "Science and Technology for Sustainable Development Special Feature: Sustainability." *Proceedings of the National Academy of Sciences (USA)* 100:8059–61.

Clement, C. R. 1999. "1492 and the Loss of Amazonian Crop Genetic Resources. 1. The Relation between Domestication and Human Population Decline." *Economic Botany* 53:188–202.

———. 2006. "Demand for Two Classes of Traditional Agroecological Knowledge in Modern Amazonia." In D. A. Pose and M. J. Balick, eds., *Human Impacts on Amazonia: The Role of Traditional Ecological Knowledge in Conservation and Development.* New York: Columbia University Press, 33–50.

Clements, F. E. 1916. *Plant Succession: An Analysis of the Development of Vegetation.* Washington, DC: Carnegie Institute of Washington (Publ. 242).

Clewell, A. F. 1995. "Downshifting." *Restoration and Management Notes* 13:171–75.

———. 1999. "Restoration of Riverine Forest at Hall Branch on Phosphate-Mined Land, Florida." *Restoration Ecology* 7:1–14.

———. 2000a. "Restoring for Natural Authenticity." *Ecological Restoration* 18:216–17.

———. 2000b. "Editorial: Restoration of Natural Capital." *Restoration Ecology* 8:1.

———. 2001. "Resistance to Restoration." *Ecological Restoration* 19:3–4.

———. 2011. "Forest Succession after 43 Years Without Disturbance on Ex-arable Land, Northern Florida." *Castanea* 76:386–94.

Clewell, A. F., and J. Aronson. 2006. "Motivations for the Restoration of Ecosystems." *Conservation Biology* 20:420–28.

———. 2007. *Ecological Restoration: Principles, Values and Structure of an Emerging Profession.* Washington, DC: Island Press.

Clewell, A. F., J. A. Goolsby, and A.G. Shuey. 1982. "Riverine Forests of the South Prong Alafia River System, Florida." *Wetlands* 2:21–72.

Clewell, A. F., J. P. Kelly, and C. L. Coultas. 2000. "Forest Restoration at Dogleg Branch on Phosphate-Mined and Reclaimed Land, Florida." In W. L. Daniels and S. G. Richardson, eds., *Proceedings, 2000 Annual Meeting of the American Society for Surface Mining and Reclamation, Tampa, Florida, June 11–15.* Lexington: American Society of Surface Mining, 197–204.

Clewell, A. F., and R. Lea. 1990. "Creation and Restoration of Forested Wetland Vegetation in the Southeastern United States." In J. A. Kusler and M. E. Kentula, eds.,

*Wetland Creation and Restoration: The Status of the Science.* Washington, DC: Island Press, 195–231.

Clewell, A., and T. McDonald. 2009. "Relevance of Natural Recovery to Ecological Restoration." *Ecological Restoration* 27:122–24.

Clewell, A. F., J. Rieger, and J. Munro. 2005. *Guidelines for Developing and Managing Ecological Restoration Projects, 2nd ed.* http://www.ser.org/ and Tucson: Society for Ecological Restoration International.

Clewell, A., and J. D. Tobe. 2011. *Cinnamomum-Ardisia* Forest in Northern Florida. *Castanea* 76:245–54.

Cole, R. J., K. D. Holl, and R. A. Zahawi. 2010. "Seed Rain under Tree Islands Planted to Restore Degraded Lands in a Tropical Agricultural Landscape." *Ecological Applications* 20:1255–69.

Costanza, R. 1992. "Toward an Operational Definition of Ecosystem Health." In R. Costanza, B. G. Norton, and B. D. Haskell, eds., *Ecosystem Health: New Goals for Environmental Management.* Washington, DC: Island Press, 239–56.

Costanza, R., and H. E. Daly. 1992. "Natural Capital and Sustainable Development." *Conservation Biology* 6:37–46.

Costanza, R., W. J. Mitsch, and J. W. Day Jr. 2006. "Creating a Sustainable and Desirable New Orleans." *Ecological Engineering* 26:317–20.

Cowles, H. C. 1899. "The Ecological Relations of the Vegetation on the Sand Dunes of Lake Michigan. 1. Geographical Relations of the Dune Floras." *Botanical Gazette* 27:95–117, 167–202, 281–308, 361–91.

Cowling, R. M., S. Proches, and J. H. J. Vlok. 2005. "On the Origin of Southern African Subtropical Thicket Vegetation." *South African Journal of Botany* 71:1–23.

Cox, A. C., D. R. Gordon, J. L. Slapcinsky, and G. S. Seamon. 2004. "Understory Restoration in Longleaf Pine Sandhills." *Natural Areas Journal* 24:4–14.

Craft, C. B., S. W. Broome, and C. L. Campbell. 2002. "Fifteen Years of Vegetation and Soil Development Following Brackish-water Marsh Creation." *Restoration Ecology* 10:248–58.

Crosti, R., P. G., Ladd, K. W. Dixon, and B. Piotto. 2006. "Post-fire Germination: The Effect of Smoke on Seeds of Selected Species from the Central Mediterranean Basin." *Forest Ecology and Management* 221:306–12.

Crutzen, P. J. 2002. "Geology of Mankind." *Nature* 415:23.

Crutzen, P. J., and E. F. Stoermer. 2000. "The 'Anthropocene.'" *Global Change Newsletter* 41:12–13.

Daily, G., ed. 1997. *Nature's Services.* Washington, DC: Island Press.

Daly, H. E., and J. B. Cobb Jr. 1994. *For the Common Good: Redirecting the Economy toward Community, the Environment, and a Sustainable Future.* 2nd ed. Boston: Beacon Press.

Daly, H. E., and J. Farley. 2004. *Ecological Economics: Principles and Applications.* Washington, DC: Island Press.

D'Antonio, C. M., and J. C. Chambers. 2006. "Using Ecological Theory to Manage or Restore Ecosystems Affected by Invasive Plant Species." In D. A. Falk, M. A. Palmer, and J. B. Zedler, *eds., Foundations of Restoration Ecology*. Washington, DC: Island Press, 260–79.

Davis, M. A., M. K. Chew, R. J. Hobbs, A. E. Lugo, J. J. Ewel, G. J. Vermeij, J. H. Brown et al. 2011. "Don't Judge Species on Their Origins." *Nature* 474:153–54.

Davis, M. A., and L. B. Slobodkin. 2004a. "The Science and Values of Restoration Ecology." *Restoration Ecology* 12:1–3.

———. 2004b. "Letter." *Frontiers in Ecology and the Environment* 2:44–45.

Davis, M. B. 1976. "Pleistocene Biogeography of the Temperate Deciduous Forests." *Geoscience and Man* 13:13–26.

Day, G. M. 1953. "The Indian as an Ecological Factor in the Northeastern Forest." *Ecology* 34:329–46.

Day, J. W., J. Barras, E. Clairain, J. Johnston, D. Justic, G. P. Kemp, J. Ko et al. 2005. "Implications of Global Climatic Change and Energy Cost and Availability for the Restoration of the Mississippi Delta." *Ecological Engineering* 24:253–65.

de Groot, R. S. 1992. *Functions of Nature*. Groningen: Wolters-Noordhoff.

de Groot, R. S., B. Fisher, M. Christie, J. Aronson, L. Braat, R. Haines-Young, J. Gowdy et al. 2010. "Integrating the Ecological and Economic Dimensions in Biodiversity and Ecosystem Service Valuation." In P. Kumar, ed., *The Economics of Ecosystems and Biodiversity: Ecological and Economic Foundations*. London and Washington, DC: Earthscan, 9–40.

Denevan, W. M. 1992. "The Pristine Myth: The Landscape of the Americas in 1492." *Annals of the Association of American Geographers* 82:369–85.

Desai, N. 2003. *Sacred Grove: Potential Reference Site for Restoration Project. Restoration Ecology Promotion Booklet 2003*. Pune: SER-India Publication Series No. 4.

Descheemaeker, K., J. Nyssen, J. Poesen, D. Raes, M. Haile, B. Muys, and S. Deckers. 2006. "Runoff on Slopes with Restoring Vegetation: A Case Study from the Tigray Highlands, Ethiopia." *Journal of Hydrology* 331:219–41.

Diamond, J. 2005. *Collapse. How Societies Choose to Fail or Succeed*. New York: Penguin Books.

Diamond, J. M. 1975. "Assembly of Species Communities." In M. L. Cody, and J. M. Diamond, eds., *Ecology and Evolution of Communities*. Cambridge: Harvard University Press, 342–444.

Doyle, M., and C. A. Drew, eds. 2008. *Large-scale Ecosystem Restoration: Five Case Histories from the United States*. Washington, DC: Island Press.

Dregne H. E. 1992. Degradation and Restoration of Arid Lands. International Center for Arid and Semiarid Land Studies, Texas Tech University, Lubbock, Texas, USA.

Ecotrust. 2000. "A Watershed Assessment for the Siuslaw Basin." Unpublished report by Ecotrust, Portland Oregon. http://www.inforain.org/siuslaw/.

Egan, D., and E. A. Howell. 2001. *The Historical Ecology Handbook. A Restorationist's Guide to Reference Ecosystems*. Washington, DC: Island Press.

Egler, F. E. 1954. "Vegetation Science Concepts 1. Initial Floristic Composition, a Factor in Old-field Vegetation Development." *Vegetatio Acta Geobotanica* 4:412–17.

Eisenberg, C. 2010. *The Wolf's Tooth: Keystone Predators, Trophic Cascades, and Biodiver-*

*sity*. Washington, DC: Island Press.

Elliott, S., P. Navakitbumrung, C. Kuarak, S. Zangkum, V. Anusarnsunthorn, and D. Blakesley. 2003. "Selecting Framework Tree Species for Restoring Seasonally Dry Tropical Forests in Northern Thailand Based on Field Performances." *Forest Ecology and Management* 184:177–91.

Esbjörn-Hargens, S., and M. E. Zimmerman. 2009. *Integral Ecology: Uniting Multiple Perspectives on the Natural World*. New York: Random House.

Falk, D. A., M. A. Palmer, and J. B. Zedler. 2006. "Integrating Restoration Ecology and Ecological Theory: A Synthesis." In D. A. Falk, M. A. Palmer, and J. B. Zedler, eds. *Foundations of Restoration Ecology*. Washington DC: Island Press, 341–46.

Farina, A. 2000. "The Cultural Landscape as a Model for the Integration of Ecology and Economics." *BioScience* 50:313–20.

Feiertag, J. A., D. J. Robertson, and T. King. 1989. "Slash and Turn." *Restoration and Management Notes* 7:13–17.

Fernandes, P.H., and H. S. Botelho. 2003. "A Review of Prescribed Burning Effectiveness in Fire Hazard Reduction." *International Journal of Wildland Fire* 12:117–28.

Figueroa, E., and J. Aronson. 2006. "Linkages and Values for Protected Areas: How to Make Them Worth Conserving?" *Journal of Nature Conservation* 14:225–32.

Filippi, O., and J. Aronson. 2011a. "Plantes invasives en région méditerranéenne: quelles restrictions d'utilisation préconiser pour les jardins et les espaces verts? " *Ecologia mediterranea* 36:31–54.

———. 2011b. "Useful but Potentially Invasive Plants in the Mediterranean Region: What Restrictions Should Be Placed on Their Use in Gardens?" *BGjournal* 8:27–31.

Flannery, T. 1994. *The Future Eaters: An Ecological History of the Australasian Lands and People*. Sydney, AU: Reed New Holland.

———. 2001. *The Eternal Frontier: An Ecological History of North America and Its Peoples*. New York: Penguin.

Forman, R. T. T. 1995. *Land Mosaics: The Ecology of Landscapes and Regions*. Cambridge, UK: Cambridge University Press.

Forman, R. T. T., and M. Gordon. 1986. *Landscape Ecology*. New York: Wiley.

Fox, L. R. 1988. "Diffuse Coevolution within Complex Communities." *Ecology* 69:906–7.

Friederici, P. 2006. *Nature's Restoration: People and Places on the Frontlines of Conservation*. Washington, DC: Island Press.

Funk, J. L., E. E. Cleland, K. N. Suding, and E. S. Zavaleta. 2008. "Restoration through Reassembly: Plant Traits and Invasion Resistance." *Trends in Ecology and Evolution* 23:695–703.

Funk, J. L., and S. McDaniel. 2010. "Altering Light Availability to Restore Invaded Forest: The Predictive Role of Plant Traits." *Restoration Ecology* 18:865–72.

Gadgil, M. 1995. "Prudence and Profligacy: A Human Ecological Perspective."In T. M. Swanson, ed., *The Economics and Ecology of Biodiversity Decline*. New York: Cambridge University Press, 99-110.

Gardner, J. H., and D. T. Bell. 2007. Bauxite Mining Restoration by Alcoa World Alumina Australia in Western Australia: Social, Political, Historical, and Environmental Contexts." *Restoration Ecology* 15:S3–S10.

Gleason, H. A. 1939. "The Individualistic Concept of the Plant Association." *American Midland Naturalist* 21:92–110.

Gondard, H., S. Jauffret, J. Aronson, and S. Lavorel. 2003. "Plant Functional Types: A Promising Tool for Management and Restoration of Degraded Lands." *Applied Vegetation Science* 6:223–34.

Goosem, S. P., and N. I. J. Tucker. 1995. *Repairing the Rainforest Theory and Practice of Rainforest Reestablishment in North Queensland's Wet Tropics.* Cairns: Wet Tropics Management Authority.

GPFLR (Global Partnership for Forest Landscape Restoration). 2012. Accessed February 28, 2012. http://www.inforesources.ch/pdf/focus_2_05_e.pdf.

Graham, R. W., and E. Lundelius Jr. 1984. "Coevolutionary Disequilibrium and Pleistocene Extinctions." In P. S. Martin and R. G. Klein, eds., *Quatenary Extinctions.* Tucson: University of Arizona Press, 223–49.

Grant, C. D. 2006. "State-and-Transition Successional Model for Bauxite Mining Rehabilitation in the Jarrah Forest of Western Australia." *Restoration Ecology* 14:28–37.

Grime, J. P. 1974. "Vegetation Classification by Reference to Strategies." *Nature* 250:26–31.

———. 1977. "Evidence for the Existence of Three Primary Strategies in Plants and Its Relevance to Ecological and Evolutionary Theory." *American Naturalist* 111:1169–94.

———. 1979. *Plant Strategies and Vegetation Processes.* Chichester, UK: Wiley.

———. 2006. "Trait Convergence and Trait Divergence in Herbaceous Plant Communities: Mechanisms and Consequences." *Journal of Vegetation Science* 17:255–60.

Grumbine, R. E. 1994. "What is Ecosystem Management?" *Conservation Biology* 8:27–38.

Gunderson, L. H. 2000. "Ecological Resilience—In Theory and Application." *Annual Review of Ecology and Systematics* 31:425–39.

Guralnick, L. J., P. A. Rorabaugh, and Z. Hanscom. 1984a. "Influence of Photoperiod and Leaf Age on Crassulacean Acid Metabolism in *Portulacaria afra* (L.) Jacq." *Plant Physiology* 75:454–57.

———. 1984b. "Seasonal Shifts of Photosynthesis in *Portulacaria afra* (L) Jacq." *Plant Physiology* 76:643–46.

Harper, J. L. 1987. "The Heuristic Value of Ecological Restoration." In W. R. Jordan III, M. E. Gilpin and J. D. Aber, eds., *Restoration Ecology a Synthetic Approach to Ecological Research.* Cambridge, UK: Cambridge University Press, 35–45.

Harris, J. A., and R. van Diggelen. 2006. "Ecological Restoration as a Project for Global Society." In J. van Andel and J. Aronson, eds., *Restoration Ecology: The New Frontier.* Oxford, UK: Blackwell Science, 3–15.

Henry F., P. Talon, and T. Dutoit. 2010. "The Age and History of the French Mediterranean Steppe Revisited by Soil Wood Charcoal Analysis." *The Holocene* 20:25–34.

Hey, D. L., and N. S. Philippi. 1999. *A Case for Wetland Restoration.* New York: Wiley.

Higgs, E. 2012. "History, Novelty, and Virtue in Ecological Restoration." In A. Thompson and J. Bendik-Keymer, eds., *Ethical Adaptation to Climate Change.* Cambridge, MA:

MIT Press, 81–101.

Higgs, E. S. 1997. "What is Good Ecological Restoration?" *Conservation Biology* 11:338–48.

Hobbs, R. J. 2002. "The Ecological Context: A Landscape Perspective." In M. Perrow and A. J. Davy, eds., *Handbook of Ecological Restoration*. Cambridge, UK: Cambridge University Press, 22–45.

Hobbs, R. J., S. Arico, J. Aronson, J. S. Baron, P. Bridgewater, V. A. Cramer, P. R. Epstein, et al. 2006. "Novel Ecosystems: Theoretical and Management Aspects of the New Ecological World Order." *Global Ecology and Biogeography* 15:1–7.

Hobbs, R.J., E. Higgs, and J. Harris. 2009. Novel ecosystems: implications for conservation and restoration. *Trends in Ecology & Evolution* 24: 599-605.

Hobbs, R., A. Jentsch, and V. Temperton. 2007. "Restoration as a Process of Assembly and Succession Mediated by Disturbance." In J. Walker, R. del Moral, L. Walker-and R. Hobbs, eds. *Linking Restoration and Ecological Succession*. Springer. New York, Springer, 150-67.

Hobbs, R. J., and D. A. Norton. 1996. "Towards a Conceptual Framework for Restoration Ecology." *Restoration Ecology* 4:93–110.

Hobbs, R. J., and K. N. Suding. 2007. *New Models for Ecosystem Dynamics and Restoration*. Washington, DC: Island Press.

Hobbs, R. J., and L. R. Walker. 2007. "Old Field Succession: Development of Concepts." In V. A. Cramer and R. J. Hobbs, eds., *Old Fields: Dynamics and Restoration of Abandoned Farmland*. Washington, DC: Island Press, 17–30.

Hoffman, M. T., and R. M. Cowling. 1990. "Desertification in the Lower Sundays River Valley, South Africa." *Journal of Arid Environments* 19:105–17.

Holl, K. D. 2012. "Restoration of Tropical Forests." In J. Van Andel, and J. Aronson, eds., *Restoration Ecology: The New Frontier*. 2nd ed. Oxford, UK: Blackwell, 103–14.

Holl, K. D., M. E. Loik, and E. H. V. Lin. 2000. "Tropical Montane Forest Restoration in Costa Rica: Overcoming Barriers to Dispersal and Establishment." *Restoration Ecology* 8:339–49.

Holland, K. M. 1994. "Restoration Rituals." *Restoration and Management Notes* 12:121–25.

Holling, C. S. 1973. "Resilience and Stability of Ecological Systems." *Annual Reivew of Ecology and Systematics* 4:1–24.

Holling, C. S., and L. H. Gunderson. 2002. "Resilience and Adaptive Cycles." In L. H. Gunderson, and C. S. Holling, eds., *Panarchy, Understanding Transformations in Human and Natural Systems*. Washington, DC: Island Press, 25–62.

House, F. 1996. "Restoring Relations: The Vernacular Approach to Ecological Restoration." *Restoration and Management Notes* 14:57–61.

Hutchinson, G. E. 1959. Homage to Santa Rosalia, or Why Are There So Many Kinds of Animals? *American Midland Naturalist* 93:145–59.

Inouye, B., and J. R. Stinchcombe. 2001. "Relationships between Ecological Interaction Modifications and Diffuse Coevolution: Similarities, Differences, and Causal Links." *Oikos* 95:353–60.

Izhaki, I., N. Henig-Sever, and G. Ne'eman. 2000. "Soil Seed Banks in Mediterranean Aleppo Pine Forests: The Effect of Heat, Cover and Ash on Seedling Emergence." *Journal of Ecology* 88:667–75.

Janzen, D.H. 1988. "Tropical Ecological and Biocultural Restoration." *Science* 239:24344.

———. 1992. "The Neotropics." *Restoration and Management Notes* 10:8–13.

———. 1998. "Gardenification of Wildland Nature and the Human Footprint." *Science* 279:1312–13.

———. 2002. "Tropical Dry Forest Restoration: Area de Conservación Guanacaste, Northwestern Costa Rica." In M. R. Perrow and A. J. Davy, eds., *Handbook of Ecological Restoration*. Vol. 2. *Restoration in Practice*. Cambridge, UK: Cambridge University Press, 559—84.

Jaunatre R., E. Buisson, and T. Dutoit. 2012. "First-Year Results of a Multi-Treatment Steppe Restoration Experiment in La Crau (Provence, France)." *Plant Ecology and Evolution* 145 :13–23.

Jones, C. G., J. H. Lawton, and M. Shachak. 1994. "Organisms as Ecosystem Engineers." *Oikos* 69:373–86.

Jordan, W. R., III. 1986. "Restoration and the Reentry of Nature." *Restoration and Management Notes* 4:2.

———. 1994. "'Sunflower Forest': Ecological Restoration as the Basis for a New Environmental Paradigm. In D. A Baldwin Jr., J. DeLuce, and C. Pletsch, eds., *Beyond Preservation: Restoring and Inventing Landscapes*. Minneapolis and London, University of Minnesota Press, 17–34.

———. 2003. *The Sunflower Forest: Ecological Restoration and the New Communion with Nature*. Berkeley: University of California Press.

Jurdant, M., J. L.Bélair, V. Gerardin, and J. P. Ducroc. 1977. *L'inventaire du capital-nature. Méthode de classification et de cartographie écologique du territoire*. Québec: Service des Etudes Ecologiques Régionales, Direction Régionale des Terres, Pêches et Environnement.

Kangas, P. C. 2004. *Ecological Engineering Principles and Practice*. Boca Raton: Lewis Publishers, CRC Press.

Kaplan, J. O., K. Krumhardt, and N.E. Zimmerman. 2009. "The Prehistoric and Preindustrial Deforestation of Europe." *Quaternary Sciences Reviews* 28: 3016–34.

Kassas, M. 1995. "Desertification: A General Review." *Journal of Arid Environments* 30:115–28.

Kates, R. W., W. C. Clark, R. Corell , J. M. Hall, C. C. Jaeger, I. Lowe, J. J. McCarthy et al. 2001. "Sustainability Science." *Science* 292:641–42.

Katz, E. 2012. "Further Adventures in the Case against Restoration." *Environmental Ethics* 34:67–97.

Keeley, J. E., W. J. Bond, R. A. Bradstock, J. G. Pausas and P. W. Rundel. 2012. *Fire in Mediterranean Ecosystems. Ecology, Evolution and Management*. Cambridge, UK: Cambridge University Press.

Koch, J. M. 2007 . "Restoring a Jarrah Forest Understorey Vegetation after Bauxite Mining in Western Australia." *Restoration Ecology* 15: Supplement s4: S26–S39.

Komarek, E. V. 1966. "The Meteorological Basis for Fire Ecology." *Proceedings of the Tall Timbers Fire Ecology Conference* 5:85125.

Krauss, S., and T. Hua He. 2006. "Rapid Genetic Identification of Local Provenance Seed Collection Zones for Ecological Restoration and Biological Conservation." *Journal for Nature Conservation* 14:190–99.

Kuhn, T. S. 1996. *The Structure of Scientific Revolutions*. 3rd ed. Chicago: University of Chicago Press.

Kurz, H., and D. Wagner. 1957. "Tidal Marshes of the Gulf and Atlantic Coasts of North Florida and Charleston, South Carolina." *Florida State University Studies* 24:1–168.

Lackey, R. T. 2004. "Societal Values and the Proper Role of Restoration Ecologists." *Frontiers in Ecology and the Environment* 2:45–46.

Lamb, D., P. D. Erskine and J. D. Parrotta. 2005. "Restoration of Degraded Tropical Forest Landscapes." *Science* 310:1628–32.

Lamb, D., and D. Gilmour. 2003. *Rehabilitation and Restoration of Degraded Forests*. Gland: International Union for Conservation of Nature and World Wide Fund for Nature.

Lara, A., D. Soto, J. Armesto, P. Donoso, C. Wernli, L. Nahuelhual, and F. Squeo. 2003. "Componentes científicos clave para una política nacional sobre usos, servicios y conservación de los bosques nativos Chilenos." Valdivia: Universidad Austral de Chile.

Lara, A., C. Little, R. Urrutia, J., C. Álvarez-Garretón, C. Oyarzún, D. Soto, P. Donoso, M. Nahuelhual, M. Pino, and I. Arismendi. 2009. "Assessment of Ecosystem Services as an Opportunity for the Conservation and Management of Native Forest in Chile." *Forest Ecology and Management* 258:415–24.

Lavelle, P. 1997. "Faunal Activities and Soil Processes: Adaptive Strategies That Determine Ecosystem Function." *Advances in Ecological Research* 27:93–132.

Lavorel, S., S. McIntyre, J. Landsberg, and T. D. A. Forbes. 1997. "Plant Functional Classification: From General Groups to Specific Groups Based on Response to Disturbance." *Trends in Ecology and Evolution* 12:474–78.

Lechmere-Oertel, R. G., G. I. H Kerley, R. M. Cowling. 2005. "Landscape Dysfunction and Reduced Spatial Heterogeneity in Soil Resources and Fertility in Semi-arid Succulent Thicket, South Africa." *Austral Ecology* 30:615–24.

Lechmere-Oertel, R. G., G. I. H Kerley, A. J. Mills, and R. M. Cowling. 2008. "Litter Dynamics across Browsing-induced Fenceline Contrasts in Succulent Thicket, South Africa. *South African Journal of Botany* 74:651–59.

Leopold, A. 1949. *A Sand County Almanac*. Oxford, UK: Oxford University Press. Reprinted in 1970 by Ballantine Books, New York.

Leopold, L. B., ed. 1993. *Round River. From the Journals of Aldo Leopold*. Oxford, UK: Oxford University Press.

Levin, L. A., and P. K. Dayton. 2009. "Ecological Theory and Continental Margins: Where Shallow Meets Deep." *Trends in Ecology and Evolution* 24:606–17.

Levin, L. A., and M. Sibuet. 2012. "Understanding Continental Margin Biodiversity: A New Imperative." *Annual Review of Marine Science* 4:79–112.

Little, C. 2011. "Rol de los bosques nativos en la oferta del servicio ecosistémico provisión de agua en cuencas forestales del centro sur de Chile." PhD thesis. Facultad de Ciencias Forestales. Universidad Austral de Chile.

Little, C., and A. Lara. 2010. "Ecological Restoration for Water Yield Increase as an Ecosystem Service in Forested Watersheds of South-central Chile." *Bosque* 31:175–78.

Lindenmayer, D. B., J. Fischer, A. Felton, M. Crane, D. Michael, C. Macgregor, et al. 2008. "Novel ecosystems resulting from landscape transformation create dilemmas for modern conservation practice." *ConservationLetters* 1:129-135.

Lloyd, J. W., E. van den Berg, and A. R. Palmer. 2002. "Patterns of Transformation and Degradation in the Thicket Biome, South Africa." TERU report no: 39. Port Elizabeth, SA: University of Port Elizabeth.

Lovelock, J. E. 1991. "Gaia, a Planetary Emergent Phenomenon." In W. I. Thompson, ed., *Gaia 2 : Emergence : the New Science of Becoming*. New York: Lindisfarne Press, 30–42.

Lugo, A. E. 1978. "Stress and Ecosystems." In J. H. Thorp and J. W. Gibbons, eds., *Energy and Environmental Stress in Aquatic Ecosystems*. Springfield, Virginia: DOE Symposium Series (CONF-771114). National Technical Information Service.

MA (Millennium Ecosystem Assessment). 2005. *Ecosystems and Human Well-Being: Synthesis*. Washington, DC: Island Press.

MacArthur, R. H., and E. O. Wilson. 1967. *The Theory of Island Biogeography*. Princeton: Princeton University Press.

Maestre, F. T., J. L. Quero, N. J. Gotelli, A. Escudero et al. 2012. "Plant Species Richness and Ecosystem Multifunctionality in Blobal Drylands." *Science* 335:214–18.

Mann, C. C. 2005. *1491. New Revelations of the Americas before Columbus*. New York: Vintage Books.

———. 2011. *1493. Uncovering the New World Columbus Created*. New York: Alfred A. Knopf.

Marais, C., R. M. Cowling, M. Powell, and A. Mills. 2009. "Establishing the Platform for a Carbon Sequestration Market in South Africa: The Working for Woodlands Subtropical Thicket Restoration Programme." *8th World Forestry Congress*. Buenos Aires, October, 18–23.

Martin, R. E., R. L. Miller, and C. T. Cushwa. 1975. "Germination Response of Legume Seeds Subjected to Moist and Dry Heat." *Ecology* 56:1441–45.

Maschinski, J., and S. J. Wright. 2006. "Using Ecological Theory to Plan Restorations of the Endangered Beach Jacquemontia (*Convolvulaceae*) in Fragmented Habitats." *Journal for Nature Conservation* 14:180–89.

Maser, C., and J. R. Sedell. 1994. *From the Forest to the Sea: The Ecology of Wood in Streams, Rivers, Estuaries, and Oceans*. Delray Beach: St. Lucie Press.

McCann, J. M. 1999. "The Making of the Pre-Columbian Landscape—Part 1: The Environment." *Ecological Restoration* 17:15–30.

McCarthy, H. 1993. "Managing Oaks and the Acorn Crop." In T. C. Blackburn and K. Anderson, eds., *Before the Wilderness: Environmental Management by Native Californians*. Anthropological papers, No. 40. Barning: Ballena Press, 213–28.

McCarty, K. 1998. "Landscape-scale Restoration in Missouri Savannas and Woodlands."

*Restoration and Management Notes* 16:22–32.

McDonald, M. C. 1996. "Ecosystem Resilience and the Restoration of Damaged Plant Communities: A Discussion Focusing on Australian Case Studies." PhD thesis, University of Western Sydney-Hawksbury.

McDonald, T. 2000. "Resilience, Recovery and the Practice of Restoration." *Ecological Restoration* 18:10–20.

———. 2005. "Grassland Restoration: Strengthening Our Underpinning." *Ecological Management and Restoration* 6:2.

———. 2008. "Evolving Restoration Principles in a Changing World." *Ecological Restoration and Management* 9:165–67.

McHarg, I. L. 1967. *Design with Nature*. New York: Wiley.

McKey, D., S. Rostain, J. Iriate, B. Glaser, J. J. Birk, I. Holst, and D. Renard. 2010. "Pre-Columbian Agricultural Landscapes, Ecosystem Engineers, and Self-Organized Patchiness in Amazonia." *Proceedings of the National Academy of Sciences USA* 107:7823–28.

Mills, A. J., J. Blignaut, R. M. Cowling, A. Knipe, C. Marais, S. Marais, M. Powell, A. Sigwela, and A. L. Skowno. 2009. "Investing in Sustainability Restoring Degraded Thicket, Creating Jobs, Capturing Carbon and Earning Green Credit." Working for Woodlands: Department of Water and the Environment. http://docs.lead.org/docs/IS 2010/Thicket_restoration.pdf.

Mills, A. J., and R. M. Cowling. 2006. "Rate of Carbon Sequestration at Two Thicket Restoration Sites in the Eastern Cape, South Africa." *Restoration Ecology* 14:38–49.

———. 2010. "Below-ground Carbon Stocks in Intact and Transformed Subtropical Thicket Landscapes in Semi-arid South Africa." *Journal of Arid Environments* 74:93–100.

Mills, A. J., and M. V. Fey. 2004. "Transformation of Thicket to Savanna Reduces Soil Quality in the Eastern Cape, South Africa." *Plant and Soil* 265:153–63.

Milton S. J. 2003. "'Emerging Ecosystems'—A Washing-stone for Ecologists, Economists and Sociologists?" *South African Journal of Science* 99:404–6.

Milton, S. J., W. R. J. Dean, M. A. du Plessis, and W. R. Siegfried. 1994. "A Conceptual Model of Arid Rangeland Degradation: The Escalating Cost of Declining Productivity." *BioScience* 44:70–76.

Milton, S. J., W. R. J. Dean, and D. M. Richardson. 2003. "Economic Incentives for Restoring Natural Capital in Southern African Rangelands." *Frontiers in Ecology and the Environment* 1:247–54.

Milton, S. J., and M. T. Hoffman. 1994. "The Application of State-and-Transition to Rangeland Research and Management in Arid Succulent and Semi-arid Grassy Karoo South Africa." *African Journal of Range and Forage Science* 11:18–26.

Milton, S. J., J. Wilson, D. M. Richardson, C. Seymour, W. Dean, D. Iponga, and S. Proches. 2007. "Invasive Alien Plants Infiltrate Bird-mediated Shrub Nucleation Processes in Arid Savanna." *Journal of Ecology* 95:648–61.

Mitsch, W. J., and S. E. Jørgensen. 2004. *Ecological Engineering and Ecosystem Restoration*. New Jersey: Wiley.

Molinier, R., and G. Tallon. 1950. "La végétation de la Crau (Basse-Provence)." *Revue*

*Générale de Botanique* 56–57, 1949–50: 1–111.

Moreira, F., A. I. Queiroz, and J. Aronson. 2006. "Restoration Principles Applied to Cultural Landscapes." *Journal for Nature Conservation* 14:207–16.

Moreno-Mateos, D., and J. Aronson. In review. "Ecological Restoration to Support Structural and Functional Recovery in Wetland Ecosystems: How Much Is Enough?" *Ecology Letters.*

Moreno-Mateos, D., M. E. Power, F. A. Comin, and R. Yockteng. 2012. "Structural and Functional Loss in Restored Wetland Ecosystems." *PLoS Biology* (10)1:e1001247.

Morrison, M. 2010. *Wildlife Restoration: Ecological Concepts and Practical Applications.* 2nd ed. Washington, DC: Island Press.

Munro, J. W. 1991. Wetland Restoration in the Mitigation Context. *Restoration & Management Notes* 9:80–86.

———. 2006. "Ecological Restoration and Other Conservation Practices: The Difference." *Ecological Restoration* 24:182–89.

Naeem, S. 1998. "Species Redundancy and Ecosystem Reliability." *Conservation Biology* 12:39–45.

Naeem, S., and S. Li 1997. Biodiversity Enhances Ecosystem Reliability. *Nature* 390: 507–9.

Nahuelhual, L., P. Donoso, A. Lara, D., Núñez, C. Oyarzún, and E. Neira. 2007. "Valuing Ecosystem Services of Chilean Temperate Rainforests." *Environment, Development and Sustainability* 9:481–99.

Naveh, Z. 2000. "The Total Human Ecosystem: Integrating Ecology and Economics." *BioScience* 50:357–61.

Nesmith, J. C. B., A. C. Caprio, A. H. Pfaff, T. W. McGinnis, and J. E. Keeley. 2011. "A Comparison of Effects from Prescribed Fires and Wildfires Managed for Resource Bbjectives in Sequoia and Kings Canyon National Parks." *Forest Ecology and Management* 261:1275–82.

Nevle, R. J., and D. K. Bird. 2008. "Effects of Syn-pandemic Fire Reduction and Reforestation in the Tropical Americas on Atmospheric CO2 during European Conquest." *Palaeogeography, Palaeoclimatology, Palaeoecology* 264:25–38.

Nevle, R. J., D. K. Bird, W. F. Ruddiman, and R. A. Dull. 2011. "Neotropical Human-Landscape Interactions, Fire, and Atmospheric $CO_2$ during European Conquest." The Holocene 21: 853-864.

Nichols, O. G., and F. M. Nichols 2003. "Long-term Trends in Faunal Recolonization after Bauxite Mining in the Jarrah Forest of Southwestern Australia." *Restoration Ecology* 11:261–72.

Noble, J., N. MacLeod, and G. Griffin. 1997. "The Rehabilitation of Landscape Function in Rangelands." In J. Ludwig, D. Tongway, D. Freudenberger, J. Nobl, and K. Hodgkinson, eds., *Landscape Ecology, Function and Management: Principles from Australia's Rangelands.* Collingwood: CSIRO, 107–120.

Oliver, C. D., and B. C. Larson. 1990. *Forest Stand Dynamics.* New York: McGraw Hill.

Olson, D., and E. Dinerstein. 1998. "The Global 200: A Representation Approach to Conserving the Earth's Most Biologically Valuable Ecoregions." *Conservation Biology*

12:502–15.

Orr, D. W. 1994. *Earth in Mind*. Washington, DC: Island Press.

———. 2002. *The Nature of Design. Ecology, Culture, and Human Intervention*. New York: Oxford University Press.

Osenberg, C. W., B. M. Bolker, J.-S. S. White, C. M. St. Mary, and J. S. Shima. 2006. "Statistical Issues and Study Design in Ecological Restorations: Lessons Learned from Marine Reserves." In D. A. Falk, M. A. Palmer, and J. B. Zedler, eds., *Foundations of Restoration Ecology*. Washington, DC: Island Press, 280–302.

Packard, S. 1988. "Just a Few Oddball Species: Restoration and the Rediscovery of the Tallgrass Savanna." *Restoration and Management Notes* 6:13–20.

———. 1993. "Restoring Oak Ecosystems." *Restoration and Management Notes* 11:5–16.

Packard, S., and C. F. Mutel, eds. 1977. *The Tallgrass Restoration Handbook for Prairies, Savannas, and Woodlands*. Washington DC: Island Press.

Paine, R. T. 1966. "Food Web Complexity and Species Diversity." *American Naturalist* 100:65–75.

———. 1969. "A Note on Trophic Complexity and Community Stability." *American Naturalist* 103:91–93.

Palmer, M. A., R. F. Ambrose, and N. LeRoy Poff. 1997. "Ecological Theory and Community." *Restoration Ecology* 5:291–300.

Pausas, J. G., and J. E. Keeley. 2009. "A Burning Story: The Role of Fire in the History of Life." *BioScience* 59:593–601.

Pausas, J. G., J. S. Pereira, and J. Aronson. 2009. "The Tree." In J. Aronson, J. S. Pereira, and J. Pausas, eds., *Cork Oak Woodlands on the Edge: Ecology, Biogeography, and Restoration of an Ancient Mediterranean Ecosystem*. Washington, DC: Island Press, 11–21.

Pickett, S. T. A., and P. S. White. 1985. *The Ecology of Natural Disturbance and Patch Dynamics*. San Diego: Academic Press.

Pimm, S. 1991. *The Balance of Nature? Ecological Issues in the Conservation of Species and Communities*. Chicago: University of Chicago Press.

Prach, K., S. Bartha, C. B. Joyce, P. Pyšek, R. van Diggelen, and G. Wiegleb. 2001. "The Role of Spontaneous Vegetation Succession in Ecosystem Restoration: A Perspective." *Applied Vegetation Science* 4:111–15.

Prach, K., and P. Pyšek. 1994. "Spontaneous Establishment of Woody Plants in Central European Derelict Sites and Their Potential for Reclamation." *Restoration Ecology* 2:190–197.

Prach, K., P. Pyšek, and V. Jarošík. 2007. "Climate and pH as Determinants of Vegetation Succession in Central European Man-made Habitats." *Journal of Vegetation Science* 18:701–10.

Price, A. R. G., M. C. Donlan, C. R. C. Sheppard, and M. Munawar. 2012. "Environmental Rejuvenation of the Gulf by Compensation and Restoration." *Aquatic Ecosystem Health and Management* 15:5–11.

Prober, S. M., and K. R. Thiele. 2005. "Restoring Australia's Temperate Grasslands and Grassy Woodlands: Integrating Function and Diversity." *Ecological Management and*

*Restoration* 6:16–27.

Puig, H., A. Fabré, M-F. Bellan, D. Lacaze, F. Villasante, and A. Ortega. 2002. "Déserts et richesse floristique : les lomas du sud péruvien, un potentiel à conserver." *Sécheresse* 13: 215–25.

Pyne, S. J. 1995. *World Fire: The Culture of Fire on Earth*. Seattle: University of Washington Press.

Pyšek, P. 1995. "On the Terminology Used in Plant Invasion Studies." In P. Pyšek, K. Prach, M. Rejmanek, and M. Wade, eds., *Plant Invasions: General Aspects and Special Problems*. Amsterdam: SPB Academic, 71–81.

Quétier, F., and S. Lavorel. 2011. "Assessing Ecological Equivalent in Biodiversity Offset Schemes: Key Issues and Solutions." *Biological Conservation* 144:2991–99.

Ramakrishnan, P. S. 1994. "Rehabilitation of Degraded Lands in India: Ecological and Social Dimensions." *Journal of Tropical Forest Science* 7:39–63.

Ramirez-Llodra E., P. A. Tyler, M.C. Baker, O. A. Bergstad, M. R. Clark, E. Escobar, L. A. Levin, L. Menot, A. A. Rowden, C. R. Smith, C. L. Van Dover. 2011. "Man and the Last Great Wilderness: Human Impact on the Deep Sea." *PLoS ONE* 6:e22588. doi:10.1371/journal.pone.0022588.

Rapport, D. J., R. Costanza, and A. J. McMichael. 1998. "Assessing Ecosystem Health." *Trends in Ecology and Evolution* 13:397–402.

Rehounková, K., and K. Prach. 2008. "Spontaneous Vegetation Succession in Gravel— Sand Pits: A Potential for Restoration." *Restoration Ecology* 16:305–12.

Reiss, J., J. R. Bridle, J. M. Montoya, and G. Woodward. 2009. "Emerging Horizons in Biodiversity and Ecosystem Functioning Research." *Trends in Ecology and Evolution* 24:506–14.

Rengasamy, P. 2006. "World Salinization with Emphasis on Australia." *Journal of Experimental Botany* 57:1017–23.

Rhode, D. 2001. "Packrat Middens as a Tool for Reconstructing Historic Ecosystems." In D. Egan and E. A. Howell, eds., *The Historical Ecology Handbook: A Restorationist's Guide to Reference Ecosystems*. Washington, DC: Island Press, 257–93.

Richardson, D. M. (ed.) 2011. *Fifty years of invasion ecology: the legacy of Charles Elton*. Oxford: Blackwell Publishing.

Richardson D. M., F. D. Pysek, M. Rejmánek, M. G. Barbour, F. D. Panetta, and C. J. West. 2000. "Naturalization and Invasion of Alien Plants: Concepts and Definitions." *Diversity and Distribution* 6:93–107.

Rietbergen-McCracken, J., S. Macinnis, and A. Sarre. 2008. *The Forest Landscape Restoration Handbook*. London and Washington: Earthscan.

Roberts, C. M. 2002. "Deep Impact: The Rising Toll of Fishing in the Deep Sea." *Trends in Ecology and Evolution* 17:242–45.

Roberts, C.M. 2009. *The Unnatural History of the Sea*. Washington, DC: Island Press.

Rodrigues, R. R., S. Gandolfi, A. G. Nave, J. Aronson, T. E. Barreto, C. Yuri Vidal, P. H. S. Brancalion. 2010. "Large-scale Ecological Restoration of High Diversity Tropical Forests in SE Brazil." *Forest Ecology and Management* 261: 1605–13.

Rogers-Martinez, D. 1992. "The Sinkyone Intertribal Park Project." *Restoration and Man-*

*agement Notes* 10:64–69.

Rokich, D. P., K. W. Dixon, K. Sivasithamparam, and K. A. Meney. 2002. "Smoke, Mulch, and Seed Broadcasting Effects on a Woodland Restoration in Western Australia." *Restoration Ecology* 10:185–94.

Román-Dañobeytia, F., S. Levy-Tacher, J. Aronson, R. Ribeiro Rodrigues, and D. Douterlunghe. 2011. "Classification of Tropical Tree Species into Ecological Groups: Implications for Forest Management and Restoration in the Lacandon Region, Chiapas, México." *Restoration Ecology* 19. Early online version: http://onlinelibrary.wiley.com/doi/10.1111/j.1526-100X.2011.00779.x/abstract.

Römermann C., T. Dutoit, P. Poschlod, and E. Buisson. 2005. "Influence of Former Cultivation on the Unique Mediterranean Steppe of France and Consequences for Conservation Management." *Biological Conservation* 121:21–33.

Rosemund, A. D., and C. B. Anderson. 2003. "Engineering Role Models: Do Non-human Species Have the Answers?" *Ecological Engineering* 20:379–87.

Rosenfeld, J. S. 2002. "Functional Redundancy in Ecology and Conservation." *Oikos* 98:156–62.

Ross, N. J. 2011. "Modern Tree Species Composition Reflects Ancient Maya 'Forest Gardens' in Northwest Belize." *Ecological Applications* 21:75–84.

Rouse, W. H. D. 1956. *Great Dialogues of Plato*. New York: Signet Classics.

Ruddiman, W. F. 2003. "The Anthropogenic Greenhouse Era Began Thousands of Years Ago." *Climatic Change* 61:261–93.

Ruddiman, W. F., and E. C. Ellis. 2009. Effect of per-capita land use changes on Holocene forest clearance and CO2 emissions. *Quaternary Science Reviews* 28:3011–15.

Rundel, P., M. O. Dillon, B. Palma et al. 1991. "The Phytogeography and Ecology of the Coastal Atacama and Peruvian Deserts." *Aliso* 13:1–49.

Sampaio, G., C. Nobre, M. H. Costa, P. Satyamurty, B. S. Soares-Filho, and M. Cardoso. 2007. "Regional Climate Change over Eastern Amazonia Caused by Pasture and Soybean Cropland Expansion." *Geophysical Research Letters* 34:L17709. doi:10.1029/2007GL030612.

Sanderson, E. W., M. Jaiteh, M. A. Levy, K. H. Redfrod, A. V. Wannebo, and G. Woolmer. 2002. "The Human Footprint and the Last of the Wild." *BioScience* 52:891–904.

Saunders, D., R. J. Hobbs, and P. R. Ehrlich, eds. 1993. *Nature Conservation: The Reconstruction of Fragmented Ecosystems*. Chipping Norton: Surrey Beaty.

Schlaepfer, M. A., D. F. Sax, and J. D. Olden. 2011. "The Potential Conservation Value of Non-native Species. *Conservation Biology* 25:428–37.

Schmitz, O., V. Krivan, and O. Ovadia. 2004. "Trophic Cascades: The Primacy of Trait-mediated Indirect Interactions." *Ecology Letters* 5:153–63.

Schneider, E., and J. Kay. 1994. "Life as a Manifestation of the Second Law of Thermodynamics." *Mathematical and Computer Modeling* 19:25–48.

Seamon, G. S. 1998. "A Longleaf Pine Sandhill Restoration in Northwest Florida." *Restoration & Management Notes* 16:46–50.

SER (Society for Ecological Restoration, Science and Policy Working Group). 2004. *The SER Primer on Ecological Restoration*. http://www.ser.org/.

Sheldrake, F. 1981. *A New Science of Life. The Hypothesis of Formative Causation*. Los Angeles: J. P. Rarcher.

Sheldrake, R. 2012. *The Science Delusion: Freeing the Spirit of Inquiry*. London: Coronet.

Shono, K., E. A. Cadaweng, and P. B. Durst. 2007. "Application of Assisted Natural Regeneration to Restore Degraded Tropical Forestlands." *Restoration Ecology* 15:620–26.

Shore, D. 1997. "The Chicago Wilderness and Its Critics 2: Controversy Erupts over Restoration in Chicago Area." *Restoration and Management Notes*: 15:25–31.

Sigwela, A. M., G. I. H. Kerley, A. J. Mills, and R. M. Cowling. 2009. "The Impact of Browsing-induced Degradation on the Reproduction of Subtropical Thicket Canopy Shrubs and Trees." *South African Journal of Botany* 75: 262–67.

Sikka, A. K., J. S. Samra, V. N. Sharda, P. Samraj, and V. Lakshmanan. 2003. "Low Flow and High Flow Responses to Converting Natural Grassland into Bluegum *(Eucalyptus globulosus)* in Nilgiris Watersheds, South India." *Journal of Hydrology* 270:12–26.

Silverton, J., M. Franco, and K. McConway. 1992. "A Demographic Interpretation of Grime's Triangle." *Functional Ecology* 6:130–36.

Simberloff, D. 2011. "Non-natives. 141 Scientists Object." *Nature* 475: 36. doi: 10.1038/475036a.

Simberloff, D., and T. Dayan. 1991. "The Guild Concept and the Structure of Ecological Communities." *Annual Review of Ecology and Systematics* 22:115– 43.

Simberloff, D., J-L Martin, P. Genovesi, V. Maris, D. A. Wardle, J. Aronson, F. Courchamp, B. Galil, E. García-Berthou, M. Pascal, et al. 2012. "Biological Invasions: What's What and the Way Forward." *Trends in Ecology & Evolution*. http://dx.doi.org/10.1016/j.tree.2012.07.013 1–9.

Simberloff, D., L. Souza, M. Nuñez, N. Garcia-Barrios, and W. Bunn. Forthcoming. "The Natives are Restless, But Not Often and Mostly When Disturbed." *Ecology*.

Smith, K. D. 1994. "Ethical and Ecological Standards for Restoring Upland Oak Forests (Ohio)." *Restoration and Management Notes* 12:192–93.

Soulé, M. E., J. A. Estes, B. Miller, and D. L. Honnold. 2005. "Strongly Interactive Species: Conservation Policy, Management, and Ethics." *BioScience* 55:168–76.

Souza, F. M., and J. L. F. Batista. 2004. "Restoration of Seasonal Semideciduous Forests in Brazil: Influence of Age and Restoration Design on Forest Structure." *Forest Ecology and Management* 191:185–200.

Steenkamp, Y., B. Van Wyk, J. Victor, D. Hoare, G. Smith, T. Dold, and R. Cowling. 2005. "Maputaland-Pondoland-Albany." In R. A. Mittermeier, P. Robles-Gil, M. Hoffman, J. Pilgrim, T. Brooks, C. Goettsch Mittermeier, J. Lamoreux, and G.A.B. da Fonseca, eds., *Hotspots Revisited: Earth's Biologically Richest and Most Threatened Terrestrial Ecoregions*. Washington, DC: Conservation International, 219–29.

Stevens, W. K. 1995. *Miracle under the Oaks*. New York: Pocket Books.

Strauss, S. Y., H. Sahli, and J. K. Conner. 2005. "Toward a More Trait-centered Approach to Diffuse (Co)evolution." *New Phytologist* 165:81–90.

Stritch, L. 1990. "Landscape-scale Restoration of Barrens-Woodland within the Oak-Hickory Mosaic." *Restoration and Management Notes* 8:73–77.

Stromberg, M., C. M. D'Antonio, T. P. Young, J. Wirka, and P. R. Kephart. 2007. "California Grassland Restoration." In M. Stromberg, J. D. Corbin, and C. M. D'Antonio, eds., *Ecology and Management of California Grasslands*. Berkeley: University of California Press, 254-80.

Suding, K. N., and K. L. Gross. 2006. "The Dynamic Nature of Ecological Systems: Multiple States and Restoration Trajectories." In D. A. Falk, M. A. Palmer, and J. B. Zedler, eds., *Foundations of Restoration Ecology*. Washington, DC: Island Press, 190–209.

Suding, K. N., and R. J. Hobbs 2009. "Threshold Models in Restoration and Conservation: A Developing Framework." *Trends in Ecology & Evolution* 24: 233–88.

Sundstrom, S. P., C. R. Allen, and C. Barichievy 2012. "Species, Functional Groups, and Thresholds in Ecological Resilience." *Conservation Biology* 26:305–14.

Swanson, L. J., and A. G. Shuey. 1980. "Freshwater Marsh Reclamation in West Central Florida." In D. P. Cole, ed. *Proceedings of the Seventh Annual Conference on the Restoration and Creation of Wetlands*. Tampa: Hillsborough Community College, 51–61.

Swetnam, T. W., C. D. Allen, and J. L. Betancourt. 1999. "Applied Historical Ecology: Using the Past to Manage for the Future." *Ecological Applications* 9:1189–1206.

Tansley, A. G. 1935. "The Use and Abuse of Vegetational Concepts and Terms." *Ecology* 16:284–307.

Taylor, A. H, and A. E. Scholl. 2012. "Climatic and Human Influences on Fire Regimes in Mixed Conifer Forests in Yosemite National Park, USA." *Forest Ecology and Management* 267:144–56.

TEEB (The Economics of Ecosystems and Biodiversity). 2010. *Ecological and Economic Foundations*, P. Kumar, ed. London and Washington, DC: Earthscan.

Temperton, V. M., R. J. Hobbs, T. Nuttle, and S. Halle, eds. 2004. *Assembly Rules and Restoration Ecology*. Washington, DC: Island Press.

Terborgh, J., C. van Schaik, L. Davenport, and M. Rao, eds. 2002. *Making Parks Work: Strategies for Preserving Tropical Nature*. Washington, DC: Island Press.

Todd, N. J. 2005. *A Safe and Sustainable World: The Promise of Ecological Design*. Washington, DC: Island Press.

Tongway, D. J., and Ludwig, J. A. 2011. *Restoring Disturbed Landscapes: Putting Principles into Practice*. Washington, DC: Island Press.

Trousdell, K. B., and M. D. Hoover. 1955. "A Change in Ground-water Level after Clearcutting of Loblolly Pine in the Coastal Plain." *Journal of Forestry* 53:493–98.

UNFCCC (United Nations Framework of the Convention on Climate Change). 2001. "UNFCCC Workshop on Definitions and Modalities for Including Afforestation and Reforestation Activities under Article 12 of the Kyoto Protocol. Orvieto, Italy." April 7–9, 2002. Accessed March 28, 2012. http://unfccc.int/meetings/workshops/other_meetings/items/1082.php.

Urbanska, K. M. 1997. "Restoration Ecology Research above the Timberline: Colonization of Safety Islands on a Machine-graded Alpine Ski Run." *Biodiversity and Conservation* 6:1655–70.

Valéry, L. H., H. Fritz, J. Lefeuvre, and D. Simberloff. 2008. "In Search of a Real Definition of the Biological Invasion Phenomenon Itself." *Biological Invasions* 10:1345–51.

——. 2009. "Invasive Species Can Also Be Native." *Trends in Ecology and Evolution*

24:585.

van Aarde, R. J., A.-M. Smit, and A. S. Claassens. 1998. "Soil Characteristics of Rehabilitating and Unmined Coastal Dunes at Richards Bay, KwaZulu-Natal, South Africa." *Restoration Ecology* 6:102–10.

van Andel, J., and J. Aronson, eds. 2012. *Restoration Ecology: The New Frontier*. 2nd ed. Oxford, UK: Blackwell.

van der Maarel, E., and M. T. Sykes. 1993. "Small-scale Plant Species Turnover in a Limestone Grassland: The Carousel Model and Some Comments on the Niche Concept." *Journal of Vegetation Science* 4:179–88.

van der Vyver, M. L., R. M. Cowling, E. E. Campbell, and M. Difford. 2012. "Active Restoration of Woody Canopy Dominants in Degraded South African Semi-arid Thicket is Neither Ecologically Nor Economically Feasible." *Applied Vegetation Science* 15: 26–34.

———. Forthcoming. "Spontaneous Return of Biodiversity in Restored Subtropical Thicket: *Portulacaria afra* as an Ecosystem Engineer." *Restoration Ecology*.

Van Lear, D. H. 2004. "Upland Oak Ecology and Management." In M. A. Spetich, ed., *Upland Oak Ecology Symposium: History, Current Conditions, and Sustainability*. General Technical Report SRS-73. U.S. Department of Agriculutre, Forest Service, Southern Research Station, Asheville, 65-71.

Vermeij, G. J. 2004. *Nature: An Economic History*. Princeton: Princeton University Press.

Vlok, J. H. J., D. I. W. Euston-Brown, and R. M. Cowling. 2003. "Acocks' Valley Bushveld 50 Years On: New Perspectives on the Delimitation, Characterisation and Origin of Thicket Vegetation." *South African Journal of Botany* 69:27–51.

Wackernagel, M., and W. E. Rees. 1996. *Our Ecological Footprint: Reducing Human Impact on the Earth*. Gabriola Island: New Society Publishers.

Wackernagel, M., N. B. Schulz, and D. Deumling. 2002. "Tracking the Ecological Overshoot of the Human Economy." *Proceedings of the National Academy of Science, USA* 99:9266–71.

Walker, B. H. 1992. "Biological Diversity and Ecological Redundancy." *Conservation Biology* 6:18–23.

Weiher, E., and P. Keddy, eds. 1999. *Ecological Assembly Rules: Perspectives, Advances, Retreats*. Cambridge, UK: Cambridge University Press.

Weinstein, M., and E. Turner, eds. 2012. *Sustainability Science: Balancing Ecology and Economy*. New York: Springer.

Wellnitz, T., and N. L. Poff. 2001. "Functional Redundancy in Heterogeneous Environments: Implications for Conservation." *Ecology Letters* 4:177–79.

Wessels, K. J., S. D. Prince, M. Carroll, and J. Malherbe. 2007. "Relevance of Rangeland Degradation in Semiarid Northeastern South Africa to the Nonequilibrium Theory." *Ecological Applications* 17:815–27.

Westman, W. E. 1977. "How Much Are Nature's Services Worth?" *Science* 197:960–964.

———. 1978. "Measuring the Inertia and Resilience of Ecosystems." *BioScience* 28:705–10.

Westoby, M., B. Walker, and I. Noy-Meir. 1989. "Opportunistic Management for Rangelands Not at Equilibrium." *Journal of Range Management* 42:266–74.

Whisenant, S. G. 1999. *Repairing Damaged Wildlands: A Process-orientated, Landscape-scale Approach*. Cambridge, UK: Cambridge University Press.

White, P. S., and A. Jentsch. 2004. "Disturbance, Succession, and Community Assembly in Terrestrial Plant Communities." In V. M. Temperton, R. Hobbs, T. Nuttle, and S. Halle, eds., *Assembly Rules and Restoration Ecology: Bridging the Gap between Theory and Practice*. Washington, DC: Island Press, 342–66.

White, P. S., and J. L. Walker. 1997. "Approximating Nature's Variation: Selecting and Using Reference Information in Restoration Ecology." *Restoration Ecology* 5:338–49.

Wilber, K. 2001. *A Theory of Everything, An Integral Vision for Business, Politics, Science and Spirituality*. Boston: Shambhala.

Willems, J. H. 2001. "Problems, Approaches, and Results in Restoration of Dutch Calcareous Grassland during the Last 30 Years." *Restoration Ecology* 9:147–54.

Woodworth, P. 2006a. "What Price Ecological Restoration?" *The Scientist* (April):39–45.

———. 2006b. "Working for Water." *The World Policy Journal* (Summer):31–43.

———. 2013. *Restoring the Future*. Chicago: Chicago University Press.

Wrangham R. W., J. H. Jones, G. Laden, D. Pilbeam, and N. L. Conklin-Brittain. 1999. "The Raw and the Stolen: Cooking and the Ecology of Human Origins." *Current Anthropology* 40:567–90.

Wyant, J. G., R. A. Meganck, and S. H. Ham. 1995. "A Planning and Decision-making Framework for Ecological Restoration." *Environmental Management* 6:789–96.

Wydhayagarn, C., S. Elliott, and P. Wangpakapattanawong. 2009. "Bird Communities and Seedling Recruitment in Restoring Seasonally Dry Forest Using the Framework Species Method in Northern Thailand." *New Forests* 38:81–97.

Young, T. P., J. M. Chase, and R. T. Huddleston. 2001. "Community Success and Assembly, Comparing, Contrasting and Combining Paradigms in the Context of Ecological Restoration." *Ecological Restoration* 19: 5-18.

Zahawi, R. A., and C. K. Augspurger. 2006. "Tropical Forest Restoration: Tree Islands as Recruitment Foci in Degraded Lands of Honduras." *Ecological Applications* 16:464–78.

Zedler, J. B., and R. Langis. 1991. "Comparisons of Constructed and Natural Salt Marshes of San Diego Bay." *Restoration and Management Notes* 9:21–25.

## Authors

**Andre F. Clewell** taught botany and ecology at Florida State University in Tallahassee for sixteen years and owned a company that specialized in restoration practice for twenty-two years. He served as president of the Society for Ecological Restoration.

**James Aronson** is a restoration ecologist at the Center of Functional and Evolutionary Ecology, of the government research network (CNRS), in Montpellier, France, and curator of restoration ecology at the Missouri Botanical Garden, USA. He is representative at large for the Society for Ecological Restoration. The authors are co-coordinators of the RNC Alliance (www.rncalliance.org).

## Guides to Virtual Field Trips

**Dean Apostol** is a landscape architect, natural resource planner, writer, and teacher who lives and works on a small farm near Portland, Oregon. With Marcia Sinclair, he edited *Restoring the Pacific Northwest: The Art and Science of Ecological Restoration in Cascadia*, published in 2006 by Island Press.

**Tanya Balcar** was born in England and lives in Kodaikanal, India. She is a founding member of the Vattakanal Conservation Trust, a conservationist, and a self-taught botanist with particular interest in the flora of the Western Ghats.

**Pedro H. S. Brancalion** is professor at the University of São Paulo, Department of Forestry, Piracicaba-São Paulo, Brazil.

**Elise Buisson** is associate professor of Ecology at the Université d'Avignon et des Pays de Vaucluse, Institut Méditerranéen de Biodiversité et d'Ecologie, Avignon, France.

**Richard M. Cowling** is professor and the head of the Restoration Research

Group, Department of Botany, Nelson Mandela Metropolitan University, Port Elizabeth, South Africa.

**Shirley Pierce Cowling** is a plant ecologist by training and works in the field of scientific communication, primarily in South Africa.

**Thierry Dutoit** is senior scientist in plant ecology, Institut Méditerranéen de Biodiversité et d'Ecologie, CNRS-IMBE, Université d'Avignon, France.

**Mauro González** is associate professor on the Facultad de Ciencias Forestales y Recursos Naturales at the Universidad Austral de Chile, Valdivia, Chile.

**Renaud Jaunatre** is a PhD candidate in restoration ecology, Université d'Avignon et des Pays de Vaucluse, Institut Méditerranéen de Biodiversité et d'Ecologie, Avignon, France.

**Antonio Lara** is professor in the Laboratorio de Dendrocronología, Instituto de Silvicultura, Facultad de Ciencias Forestales y Recursos Naturales at the Universidad Austral de Chile, Valdivia, Chile.

**Christian Little** serves on the Facultad de Ciencias at the Universidad Austral de Chile, Valdivia, Chile.

**Christo Marais** is head of operations, Natural Resource Management Programs, Department of Environmental Affairs, South Africa.

**David Printiss** directs the North Florida Conservation Program for The Nature Conservancy at the Apalachicola Bluffs and Ravines Preserve, Bristol, Florida, USA.

**Jordan Secter** runs Secter Environmental Design LLC in Portland, Oregon, USA.

**Ayanda M. Sigwela** is a restoration ecologist for South African National Parks.

**Robert Stewart** was born in England and lives in Kodaikanal, India. He is a founding member of the Vattakanal Conservation Trust.

**R. Marius van der Vyver** is a PhD candidate in the Department of Botany, at Nelson Mandela Metropolitan University, Port Elizabeth, South Africa.

## Foreword Author

**Paddy Woodworth** is a journalist and author from Dublin, Ireland, and has written more accounts of ecological restoration and the restoration of natural capital for popular consumption than any other writer. He authored *Restoring the Future*, a study of restoration projects worldwide, to be published by the University of Chicago Press in 2013.

# THE SCIENCE AND PRACTICE
# OF ECOLOGICAL RESTORATION

*Wildlife Restoration: Techniques for Habitat Analysis and Animal Monitoring*, by Michael L. Morrison

*Ecological Restoration of Southwestern Ponderosa Pine Forests*, edited by Peter Friederici, Ecological Restoration Institute at Northern Arizona University

*Ex Situ Plant Conservation: Supporting Species Survival in the Wild*, edited by Edward O. Guerrant Jr., Kayri Havens, and Mike Maunder

*Great Basin Riparian Ecosystems: Ecology, Management, and Restoration*, edited by Jeanne C. Chambers and Jerry R. Miller

*Assembly Rules and Restoration Ecology: Bridging the Gap Between Theory and Practice*, edited by Vicky M. Temperton, Richard J. Hobbs, Tim Nuttle, and Stefan Halle

*The Tallgrass Restoration Handbook: For Prairies, Savannas, and Woodlands*, edited by Stephen Packard and Cornelia F. Mutel

*The Historical Ecology Handbook: A Restorationist's Guide to Reference Ecosystems*, edited by Dave Egan and Evelyn A. Howell

*Foundations of Restoration Ecology*, edited by Donald A. Falk, Margaret A. Palmer, and Joy B. Zedler

*Restoring the Pacific Northwest: The Art and Science of Ecological Restoration in Cascadia*, edited by Dean Apostol and Marcia Sinclair

*A Guide for Desert and Dryland Restoration: New Hope for Arid Lands*, by David A. Bainbridge

*Restoring Natural Capital: Science, Business, and Practice*, edited by James Aronson, Suzanne J. Milton, and James N. Blignaut

*Old Fields: Dynamics and Restoration of Abandoned Farmland*, edited by Viki A. Cramer and Richard J. Hobbs

*Ecological Restoration: Principles, Values, and Structure of an Emerging Profession*, by Andre F. Clewell and James Aronson

*River Futures: An Integrative Scientific Approach to River Repair*, edited by Gary J. Brierley and Kirstie A. Fryirs

*Large-Scale Ecosystem Restoration: Five Case Studies from the United States*, edited by Mary Doyle and Cynthia A. Drew

*New Models for Ecosystem Dynamics and Restoration*, edited by Richard J. Hobbs and Katharine N. Suding

*Cork Oak Woodlands in Transition: Ecology, Adaptive Management, and Restoration of an Ancient Mediterranean Ecosystem*, edited by James Aronson, João S. Pereira, and Juli G. Pausas

*Restoring Wildlife: Ecological Concepts and Practical Applications*, by Michael L. Morrison

*Restoring Ecological Health to Your Land*, by Steven I. Apfelbaum and Alan W. Haney

*Restoring Disturbed Landscapes: Putting Principles into Practice*, by David J. Tongway and John A. Ludwig

*Intelligent Tinkering: Bridging the Gap between Science and Practice*, by Robert J. Cabin

*Making Nature Whole: A History of Ecological Restoration*, by William R. Jordan and George M. Lubick

*Human Dimensions of Ecological Restoration: Integrating Science, Nature, and Culture*, edited by Dave Egan, Evan E. Hjerpe, and Jesse Abrams

*Plant Reintroduction in a Changing Climate*, edited by Joyce Maschinski and Kristin E. Haskins

*Tidal Marsh Restoration: A Synthesis of Science and Management*, edited by Charles T. Roman and David M. Burdick

*Ecological Restoration: Principles, Values, and Structure of an Emerging Profession*, 2nd ed., Andre F. Clewell and James Aronson